PERGAMON INTERNATIONAL LIBRARY
of Science, Technology, Engineering and Social Studies

The 1000-volume original paperback library in aid of education, industrial training and the enjoyment of leisure

Publisher: Robert Maxwell, M.C.

Reactor Core Fuel Management

Other Titles of Interest

GRANT, P. J.
Elementary Reactor Physics

HUNT, S. E.
Fission, Fusion and the Energy Crisis

MASSIMO, L.
The Physics of High Temperature Reactors

MURRAY, R. L.
Nuclear Energy

THOMAS, A. F. and ABBEY, F.
Calculation Methods for Interacting Arrays of Fissile Material

TYROR, J. G. and VAUGHAN, R. I.
An Introduction to the Neutron Kinetics of Nuclear Power Reactors

WILLIAMS, M. M. R.
Random Processes in Nuclear Reactors

Reactor Core Fuel Management

P. SILVENNOINEN

Technical Research Centre of Finland, Helsinki

PERGAMON PRESS

OXFORD · NEW YORK · ONTARIO
SYDNEY · PARIS · FRANKFURT

UK	Pergamon Press Ltd., Headington Hill Hall, Oxford, England
U.S.A.	Pergamon Press Inc., Maxwell House, Fairview Park, Elmsford, New York 10523, U.S.A.
CANADA	Pergamon of Canada, PO. Box 9600, Don Mills M3C 2T9, Ontario, Canada
AUSTRALIA	Pergamon Press (Aust.) Pty. Ltd., 19a Boundary Street, Rushcutters Bay, N.S.W. 2011, Australia
FRANCE	Pergamon Press SARL, 24 rue des Ecoles, 75240 Paris, Cedex 05, France
WEST GERMANY	Pergamon Press GmbH, 6242 Kronberg/Taunus, Pferdstrasse 1, Frankfurt-am-Main, West Germany

First edition 1976

Library of Congress Cataloging in Publication Data

Silvennoinen, P
Reactor core fuel management.

(Pergamon international library of science, technology, engineering, and social studies)
Includes index.
1. Nuclear fuels. 2. Nuclear reactors.
I. Title.
TK9360.S48 1976 621.48'35 75-42472
ISBN 0-08-019853-8
ISBN 0-08-019852-X flexi.

Printed in Great Britain by A. Wheaton & Co.

To Tella, Teemu, Petra, and Juri

Contents

Preface

THE increasing utilization of nuclear energy for power production has created a new occupation of reactor core analysts and nuclear fuel engineers employed by the power utilities. As their tool in reactor core fuel management, these people require an access to the sizable computational machinery used to solve the planning problems. This book is intended to cover the topics associated with the planning activity, ranging from the basic physical elements to the reactor core simulation on digital computers and to the principles of fuel cycle cost optimization.

Reactor core fuel management is defined to encompass those aspects of fuel composition and loading, whether related to physics, engineering, or economic decisions, that are relevant to optimal fuel utilization within the design limits imposed on the reactor core. As far as basic reactor physics is concerned, the reader is assumed to have had an introductory course as a prerequisite, and the present treatment is only cursory. The quantitative discussion on the part of reactor physics is based largely on the neutron diffusion equation which is introduced in a number of variant forms in Chapter 2. The other introductory chapters include an account of the core heat transfer and coolant flow as well as a discussion on the reactivity behaviour of the reactor. While the reactor operation occupies minor interest in this context, a concise chapter on the topic is still included here.

Major emphasis has been devoted to the methods of reactor core analysis. Following an identification of the key variables and parameters in Chapter 6, a comprehensive treatise is presented on computer algorithms and codes used for the purposes of reactor simulation. The part related intimately to physics is mostly confined to section 7.3. The condensation is performed for the benefit of a more practically oriented reader who may lack the impetus for a detailed exploration of the physics of fuel pin lattices. I have attempted to underline that the computer code system should be modular, and in fact the modern ones are, to allow a maximal degree of flexibility and versatility. Perhaps equally importantly, the system must be a unified one where the same accuracy can be maintained throughout the course. The computational methods described are selected on this basis.

It seems that the application of optimization methods has not settled to a standardized format and the discussion on them is not very rewarding. The recipes of Chapters 9 and 10 must therefore be regarded as feasible guidelines rather than strictly recommended algorithms.

This book is mainly based on two series of lectures, one delivered as a graduate course at the Helsinki University of Technology and the other constituting a part in a training programme for utility engineers organized by the Technical Research Centre of Finland. The final writing of the text was commenced while the author was a Visiting Fulbright Scholar at the Massachusetts Institute of Technology. The contribution of these institutions is gratefully acknowledged.

Finally, I wish to extend my sincere appreciation to Messrs. K. Haule, E. Kaloinen, and E. Patrakka for the illuminating discussions at the early stages of the work.

PEKKA SILVENNOINEN

PART I

INTRODUCTORY TOPICS

CHAPTER 1

Concepts of Reactor Physics

BASIC rudiments of reactor physics are reviewed in this chapter. Because later the overall objective is confined to fuel management, the present review is succinct and it is understood that the reader is already to some extent familiar with the topic. Discussion will mainly serve as a prerequisite for the subsequent portion of the text and is included for that purpose. While the treatment is not focused around any particular reactor type in principle, one may detect implicitly that the relative weight is somewhat tuned according to the light water reactor concept.

1.1. Reactor Core

Nuclear fuel comprises at least one of the fissile uranium or plutonium isotopes and thereby provides the indispensable neutron multiplication for the self-sustaining fission process. The fuel is usually manufactured into pellets or corresponding particles to form fuel rods. These rods or pins are further assembled into fuel elements.

The isotopic fuel composition is closely associated with any given reactor type. In fact, the other physically fundamental constituents of a core are dictated by the choice of fuel. A thermal reactor contains moderator material where neutrons emitted in the fissions are slowed down to velocities more efficient for splitting new fuel atoms. In power reactors either water, heavy water or graphite is used as the moderator, depending which is most economic and compatible with the fuel composition selected.

The heat generated is removed from the reactor core by means of coolant flow through the core. Based upon heat removal capacity and other material properties, light or heavy water and helium or carbon dioxide gas are used as the coolant in thermal power reactors, whereas liquid sodium or helium gas is or will be employed in commercial fast reactors. The coolant may slightly moderate neutrons in all types of reactor, and this will indeed be the case in those thermal reactors where the moderator and coolant are one and the same. Light water reactors and natural uranium heavy water reactors represent systems of this kind.

In addition to the fuel, coolant and possible moderator, the core behaviour is influenced by a variety of control devices which can increase absorption of neutrons. Control rods, elements and poisons enter the core analysis at different stages of the subsequent discussion.

The lumping of fuel is illustrated in Fig. 1.1, where the cross-section of a typical fuel assembly is depicted.[1] In this design, control rods are clustered within the assembly and the guide thimbles are shown in the figure. The fuel pellets are contained within zircaloy cladding. The pellet outer diameter is usually about 85–90% of the cladding outer diameter.

Roughly speaking most light water reactors (LWRs) have physically rather similar fuel assemblies. Moderator-to-fuel ratio varies around 1.7 for pressurized water reactors (PWRs) and around 2.4 for boiling water reactors (BWRs).

A typical PWR core consists of up to two hundred fuel assemblies of the type described in Fig. 1.1, thus corresponding to about fifty thousand fuel pins inside a reactor vessel.

The core array is schematically shown in Fig. 1.2. At this stage the smallest entity considered is a single fuel assembly. The positions of control rod clusters are indicated as well. Needless to say, the composition of each fuel element varies significantly according to their initial state and current irradiation status.

An alternative to the reactor vessel concept is furnished by other reactor designs such as the natural uranium and other heavy water moderated reactors which involve singular pressure tubes passing through a calendria filled with the moderator. Due to separation of fuel elements into pressure tubes, the core is more heterogeneous in nature and deserves some special attention from the viewpoint of reactor core fuel management.

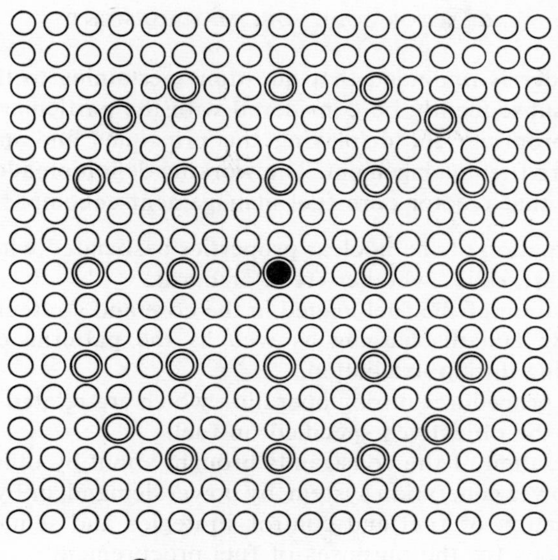

FIG. 1.1. Fuel assembly with rod cluster control.[1]

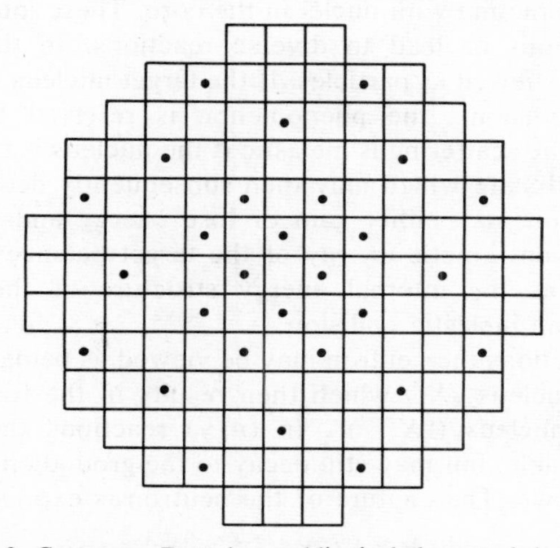

FIG. 1.2. Core array. Dotted assemblies include control elements.

The reader is reminded that questions pertaining to the reactor core will subsequently be considered with the particular emphasis found appropriate for displaying the principles and methods of reactor core fuel management. The working definition adopted here will imply that reactor core fuel management encompasses:

Determination of the nuclear properties associated with fuel and fuel assemblies whether fresh or burnt.

Specification of the fuel loaded in the core and strategies used in loading and discharging as well as the patterns of internal shuffling during irradiation.

Control procedures during operation which are related and parallel to the objectives specified on fuel.

Consideration of the constraints brought about by other units of the power generation system, by power demand, or by safety.

Optimization of fuel and control strategies and sequencing the decisions for the purposes of fuel procurement.

1.2. Neutron Interactions

Power generation in a reactor is sustained by a distribution of neutrons interacting with nuclei in the core. These interactions are either collisions or lead to diverse reactions. In the collisions, neutrons are viewed as particles. If the target nucleus is unchanged after the collisions, the phenomenon is referred to as elastic scattering. The scattering is inelastic if the nucleus is transferred to some excited state which may then subsequently decay.

The neutron may either gain or lose energy and in an elastic scattering event kinetic energy of the target balances this energy shift, whereas the internal energy structure of the nucleus is involved in an inelastic collision.

In the reactions the neutron may be viewed as being absorbed by the target nucleus ${}_ZX^A$ which then results in the formation of a compound nucleus $({}_ZX^{A+1})^*$. In (n, γ) reactions the compound nucleus is stable, but may still decay to the ground energy state by emitting γ-rays. The capture of the neutron is expressed by

$$n + {}_ZX^A \rightarrow ({}_ZX^{A+1})^* \rightarrow {}_ZX^{A+1} + \gamma. \qquad (1.1)$$

Other reactions occurring in the core are consequences of the transmutation of unstable compound nuclei. These reactions include (n, p), (n, α) and $(n, 2n)$ among which $(n, 2n)$ has a positive contribution to the neutron economy of the system. The compound nucleus may fission and remain rather fictitious. Fission can be represented in the form

$$n + {}_ZX^A \rightarrow A + B + \nu n + Q, \tag{1.2}$$

where the fission fragments A and B may decay further, yielding a number of fission products and radiation. ν denotes the average number of neutrons released in a fission and Q denotes the energy release which amounts to about 200 MeV per fission.

1.3. Neutron Flux

Because of the statistical nature of the phenomena one should think of neutrons rather as an ensemble than as individual particles. Let $N(\mathbf{r}, \mathbf{v})$ denote the expected number of neutrons at the position $\mathbf{r}$ having the velocity $\mathbf{v}$ per unit volume in the combined phase space. Adopting the nomenclature of ref. 2, this quantity is called angular density, viz. $\mathbf{v}$ will in practice be represented as a product

$$\mathbf{v} = v\mathbf{\Omega} \tag{1.3}$$

of speed v and unit direction vector $\mathbf{\Omega}$ which is two-dimensional.

A more operative quantity is the angular flux denoted by $\Phi(\mathbf{r}, \mathbf{v})$ and defined as

$$\Phi(\mathbf{r}, \mathbf{v}) = vN(\mathbf{r}, \mathbf{v}) \tag{1.4}$$

and even more so is the (total) flux $\phi(\mathbf{r}, v)$,

$$\phi(\mathbf{r}, v) = \int_{4\pi} \Phi(\mathbf{r}, \mathbf{v})\, d\mathbf{\Omega}. \tag{1.5}$$

All these quantities will vary with time, but since the time dependency will not substantially concern the subsequent discussion, the time variable is omitted here. While the direct physical meaning was attached to the angular density, the neutron flux $\phi(\mathbf{r}, v)$ can be understood to represent the number of neutrons per unit time intersecting a unit surface at $\mathbf{r}$ with the speed v.

1.4. Neutron Cross-Sections

The strength of neutron interactions with media is measured in terms of cross-sections specified separately for each kind of target atoms or molecules and for each form of interactions. The macroscopic cross-section $\Sigma_i^m(\mathbf{v})$ is the probability per unit neutron path length for a neutron in the medium m to experience an interaction of type i with the medium. Here $\mathbf{v}$ denotes the velocity of neutrons.

If the concept of cross-section is approached from the fundamental principles of quantum mechanics,[3] one customarily deals mainly with the microscopic cross-sections σ_i^m rather than the macroscopic ones. These two are associated through the simple relation

$$\Sigma_i^m = N^m \sigma_i^m \tag{1.6}$$

where N^m denotes the number density of the target nuclei.

The definition of the macroscopic cross-section implies that Σ has the dimension of inverse length (cm^{-1}) and consequently the microscopic cross-section σ has the dimension of area. The basic unit, however, is the barn (b) which is equal to 10^{-22} mm^2. Incidentally, the verbal definition of Σ given above may lead the reader to infer that for a cross-section to exist the neutrons are necessarily required to traverse with some non-zero velocity. In principle, this is not the case, and the limits of the form $\lim_{v \to 0} v\,\Sigma(\mathbf{v})$ involve exigencies[3] in a more profound discussion than is relevant for the purposes of this short introduction, or of fuel management for that matter.

For any material the set of cross-sections that must be known, i.e. measured or computed employing pertinent models of the interactions involved, consists of data for different types of scattering and absorption. These cross-sections will be denoted by Σ_s and Σ_a or by σ_s and σ_a, respectively.

Note that Σ_a consists of all reactions where incident neutrons vanish. At low energies most absorptions in the fuel, for instance, consist of radiative capture and fission, i.e. $\Sigma_a^F = \Sigma_\gamma^F + \Sigma_f^F$. Total cross-section Σ_t is the sum of the cross-sections for all types of interactions. For a homogeneous mixture of different isotope species m the cross-section is obtained as a weighted sum

$$\Sigma_i = \sum_{\substack{\text{over} \\ \text{all}\, m}} N^m \Sigma_i^m. \tag{1.7}$$

Even when a heterogeneous region is treated homogenized, this formula can be frequently employed as an approximation.

In most cases dealt with in reactor physics the quantities have weak angular dependencies and they are considered as depending on speed only. The speed of incident neutrons is equivalently expressed in terms of energy, the transformation possessing the trivial form $E = 1/2mv^2$. The unit of energy is the electron volt (eV) which is equal to 1.6×10^{-19} joule in the dimensions standardized in other fields of technology.

Cross-sections and nuclear data in general are compiled in data libraries where they are available on magnetic tapes. The graphs displayed in this book are qualitative only, but the utilization of the data compilations will be described in section 7.2. In order to facilitate the discussion, here the total microscopic cross-section of U^{238} is given in Fig. 1.3 for energies above 0.32 eV.

At low energies in the MeV region and below, $\sigma_t(E) \sim E^{-1/2}$ for U^{238}, i.e. it has a $1/v$ behaviour which is the most characteristic one and occurs for many important light nuclei as well. An almost constant plateau extends up to a few electron volts where the

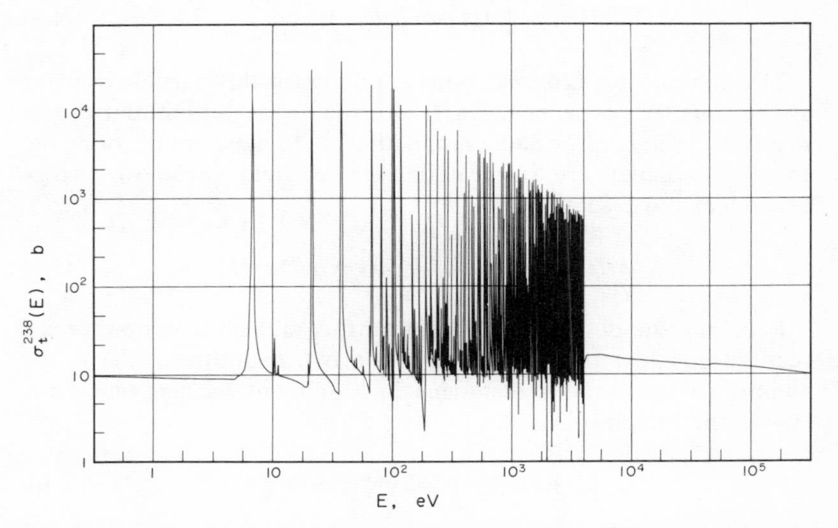

FIG. 1.3. Total cross-section as a function of energy for U^{238} (logarithmic scale, average cross-section is plotted in the unresolved range). Based on ENDF/B data.[4]

resonance region commences. This strong resonance structure and especially the unresolved resonances at higher energies are representative of the most important thorium, uranium and plutonium isotopes, whereas the lighter and intermediate elements may exhibit single isolated resonances.

Energetic neutrons in the MeV region interact less noticeably with the U^{238} atoms. Observe, however, that neutrons pass the energy scale in the reversed order from high to low energies. It is also relevant to realize that while resonances most commonly impoverish the neutron economy directly, there are fission or even scattering contributions embedded in the resonance peaks.

So far the discussion has been confined to the part of interactions where neutrons are removed from the system. In neutron preserving or producing reactions secondary neutrons are emitted and cross-section data will involve the corresponding velocity distributions. Suppose that a neutron possessing velocity $\mathbf{v}$ is scattered, then a distribution $f_s(\mathbf{v}, \mathbf{v}')$ is introduced to indicate the probability that the neutron will emerge with velocity $\mathbf{v}'$ from the collision. As a normalization one obtains readily

$$\int f(\mathbf{v}, \mathbf{v}')\, d\mathbf{v}' = 1. \tag{1.8}$$

The microscopic cross-sections counting for this particle transfer in the angular phase space are referred to as differential cross-sections in the nuclear data compilations.[4] In most applications the angular variables are integrated over to yield speed or energy dependent cross-sections

$$\sigma_s(v, v') = \int\int \sigma_s(\mathbf{v}) f_s(\mathbf{v}, \mathbf{v}')\, d\boldsymbol{\Omega}\, d\boldsymbol{\Omega}'. \tag{1.9}$$

Recalling the definitions of neutron flux and macroscopic cross-section, consider the product of these two quantities. $\Sigma_i^m \phi$ is the number of interactions occurring in a unit volume per unit time. Hence the integral

$$R_i^m(\mathbf{r}) = \int \Sigma_i^m(\mathbf{v}) \Phi(\mathbf{r}, \mathbf{v})\, d\mathbf{r} \tag{1.10}$$

corresponds to the rate of reactions taking place at $\mathbf{r}$ within medium

m where there prevails neutron flux $\Phi(\mathbf{r}, \mathbf{v})$. $R_i(\mathbf{r})$ is called the reaction rate.

1.5. Fission and Energy Release

In accordance with any other reaction, fission rate incorporates both the behaviour of the fission cross-section for the fuel composition selected and the energy variation of the neutron flux maintained in the chain-reacting system. A qualitative graph of $\sigma_f(E)$ of U^{235} is depicted in Fig. 1.4. In view of Fig. 1.4, low energy neutrons are far more amenable to induce fission than what high energy neutrons would be. However, all parasitic absorptions have to be taken into account in a due comparison. That is why fuel efficiency is more immediately given by the average number of neutrons released in fission per neutron absorbed in fuel. This ratio $\eta = \nu\sigma_f/\sigma_a$ varies

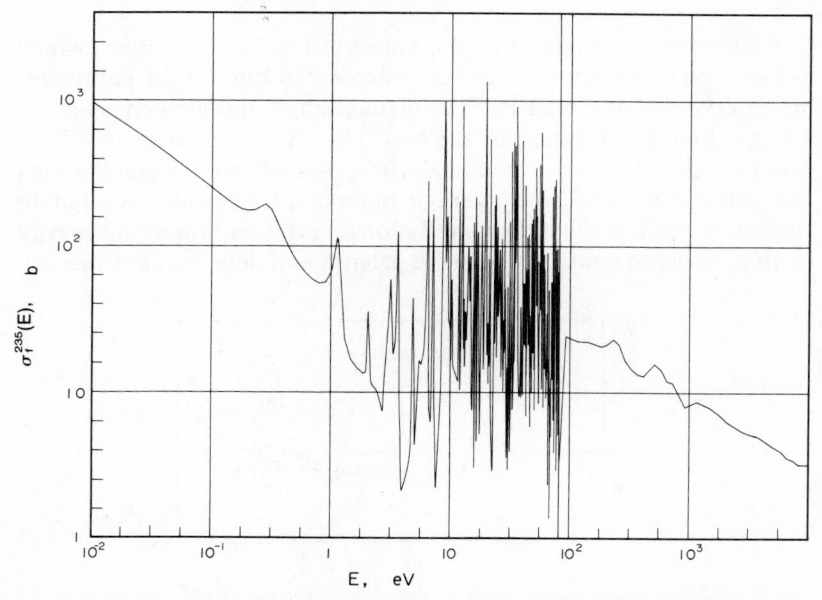

FIG. 1.4. Fission cross-section as a function of energy for U^{235} (logarithmic scale, average cross-section is plotted in the unresolved range). Based on ENDF/B data.[4]

strongly with energy for the most important fissile isotopes used as primary reactor fuels. For this and a number of other more practical reasons the major bulk of fission-inducing neutrons are designed to occupy in power reactors either the energy range below 1 eV or above 100 keV. On this basis reactors are categorized into thermal or fast reactors, respectively.

Cross-section data and parameter η are listed in Table 1.1 for fissile isotopes at energy $E = 0.0253$ eV.

TABLE 1.1
Thermal Cross-sections and η for Fissile Isotopes

Nuclide	σ_a, barn	σ_f, barn	η
U^{233}	576	528	2.29
U^{235}	680	579	2.08
Pu^{239}	1008	742	2.13
Pu^{241}	1376	1013	2.21

Within the fast energy range η varies around 2.70 for Pu^{239}, which is the main fuel composite of fast reactors. In Fig. 1.5 the parameter η is drawn for U^{235} and Pu^{239} as a function of the incident neutron energy. Figure 1.5 reiterates the fact that Pu^{239} is superior to U^{235} at fast energies. From the practical viewpoint of reactor calculations the parameter $\nu(E)$, the average number of neutrons emitted in fission, is tabulated in data compilations[4] as a function of the energy of the incoming neutron. Both the prompt and delayed neutrons are

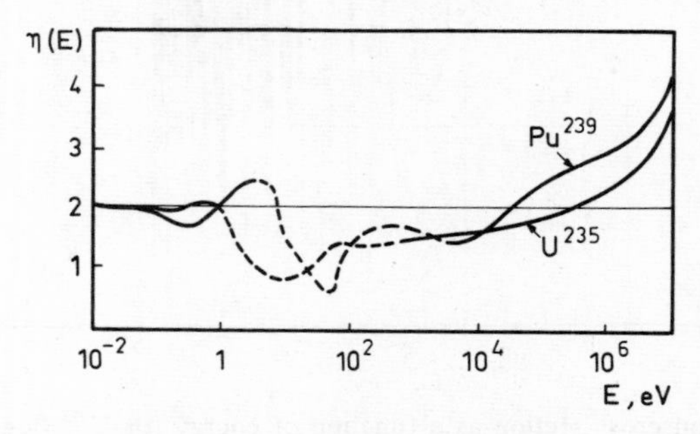

FIG. 1.5. $\eta(E)$ for U^{235} and Pu^{239}.

treated as a whole and need not be separated in the considerations of interest here. Qualitatively, $\nu(E)$ varies linearly with energy

$$\nu(E) = a + bE \tag{1.11}$$

with different constants a and b for different fissionable isotopes.

The energy distribution of prompt secondary neutrons emitted in fission, denoted usually as $\chi(E)$, is contained in data libraries as well. This so-called fission spectrum obeys qualitatively the relation

$$\chi(E) = 2\sqrt{a^3/\pi b}\,\exp\left(-\frac{ab}{4} - \frac{E}{a}\right)\sinh\sqrt{bE}. \tag{1.12}$$

Note that this expression of $\chi(E)$ implies the normalization

$$\int_0^\infty \chi(E)\,dE = 1. \tag{1.13}$$

The behaviour of $\chi(E)$ as computed from eq. (1.12) is shown in Fig. 1.6.

In addition to the neutrons emitted, the compound nucleus is split in fission into two or more other fragments which may decay further. For U^{235} and U^{238} based chains a somewhat simplified decay chart is seen in Fig. 1.7.

A complete set of relations governing all the fission products would necessitate the bookkeeping of some 200 isotopes.[4, 5] In

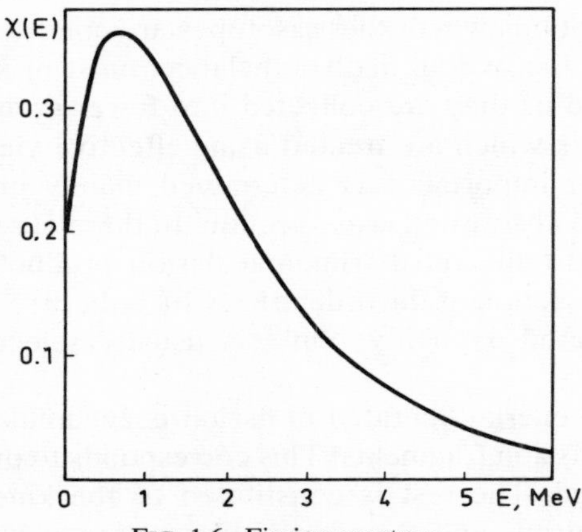

FIG. 1.6. Fission spectrum.

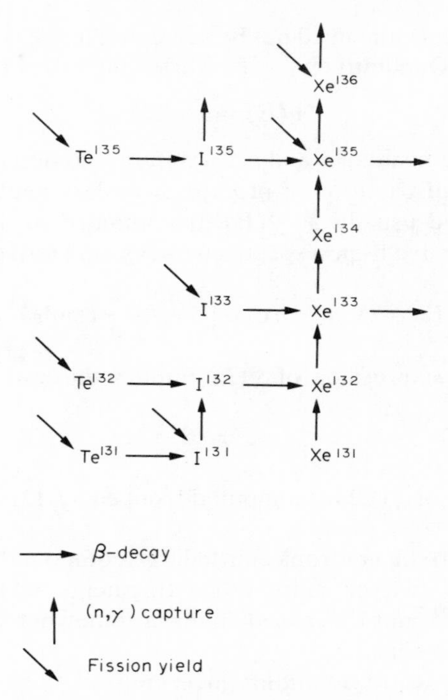

FIG. 1.7. A decay chart of fission products.

reactor calculations where these isotopes are considered within the framework of the overall neutron balance, most of single isotopes can be omitted or they are collected into fewer groups of pseudo-fission products which are treated using effective yields and cross-sections. Their importance is determined mainly on the basis of abundance and absorption cross-section. In thermal reactors Xe^{135} is the most important and detrimental fission product possessing a thermal cross-section of the order of 3×10^6 b. In case more than one isotope is treated explicitly, Sm^{149} is usually selected next after Xe^{135}.

Most of the energy liberated in fission is accumulated as kinetic energy of the fission fragments. This corresponds to over 80% of the energy released. The rest is constituted of the kinetic energy of secondary neutrons and β and γ radiation. Some 5% of energy is carried by neutrinos and is irrecoverable as heat. On the other hand, some non-fission energy is available as a result of radiative capture. This amounts to about 1–5% of fission energy. The effective energy yield per fission is given for certain isotopes in Table 1.2. A comprehensive review is given in ref. 6.

TABLE 1.2
Effective Energy Release per Fission

Nuclide	Energy, MeV
U^{235}	192.9
Pu^{239}	198.5
Pu^{241}	200.3

The data in Table 1.2 apply for thermal neutrons. In the fast energy range the situation is slightly different. In the net balance the energy of impinging neutrons may no more be negligible. For all practical purposes considered in this book the energy recovered is transformed into heat instantaneously and the spatial power distribution is identical to the local distribution of fission rate.

1.6. Fertile Isotopes

U^{235} is the only fissile isotope occurring in nature. The abundance of this isotope is 0.711% (of weight) in natural uranium and even resources of exploitable uranium are relatively scarce. The second most important fissile isotope Pu^{239} is produced by neutron capture in U^{238}. The mutation proceeds as

$$n + U^{238} \rightarrow U^{239} + \gamma \rightarrow Np^{239} + \beta \rightarrow Pu^{239} + \beta. \tag{1.14}$$

The nuclei where neutron capture leads to the formation of fissile material are designated fertile. The corresponding reaction or chain of reactions is known as conversion. Besides U^{238}, the other noteworthy fertile isotopes are Pu^{240} and Th^{232} which can be converted into Pu^{241} and U^{233}, respectively. Pu^{240} is formed in captures from Pu^{239} and Th^{232} occurs rather abundantly in nature.

The intensity of conversion is given by the conversion ratio defined as the average number of fissile nuclei created per one fissile nucleus consumed. Conversion ratios can be relevant to an entire reactor, core or any partial volume thereof, and the pointwise values are averaged accordingly.

In breeder reactors the conversion ratio exceeds unity and is usually referred to as the breeding ratio. The core is then curtained by a blanket of fertile material. While different designs and even hybrid concepts could conceivably be made to gain fissile material, fast breeders seem to be the sole practical solution. Mixed Pu–U fuel is employed there, and U^{238} is the fertile blanket material.

As far as thermal reactors are concerned, the conversion in the core is an important source of fissile material. Natural uranium reactors with high U^{238} concentration yield higher conversion ratios, about 0.85, than do reactors with enriched fuel. In the enrichment process U^{235} concentration is increased and the U^{238} abundance is diluted. A higher conversion ratio can even there be revived by introducing the U^{233} cycle and inserting thorium in the core.

Fertile isotopes exhibit an additional physical facet by being fissionable in the sense that sufficiently energetic neutrons can induce them to fission. This is indicated in Fig. 1.8 where the fission cross-section of U^{238} is displayed. It is seen that the energy of incoming neutrons shall exceed 821 keV in order to be able to cause disintegration which is now called fast fission. The amount of energy released per fast fission in U^{238} is equal to 193.9 MeV.[6] For example, in typical LWRs some 5–6% of energy output is due to fast fissions in U^{238}.

1.7. Energy Spectrum of Neutrons

The life history of neutrons in a reactor can be viewed most illustratively in the energy scale. Consider the sequence of events that follow an absorption in the fuel. In the ensuing fission η neutrons are inserted in the chain on the average. As shown in Fig. 1.6, these fission neutrons carry an amount of energy most probably of about 1 MeV while the average energy of these prompt neutrons is equal to 2 MeV.

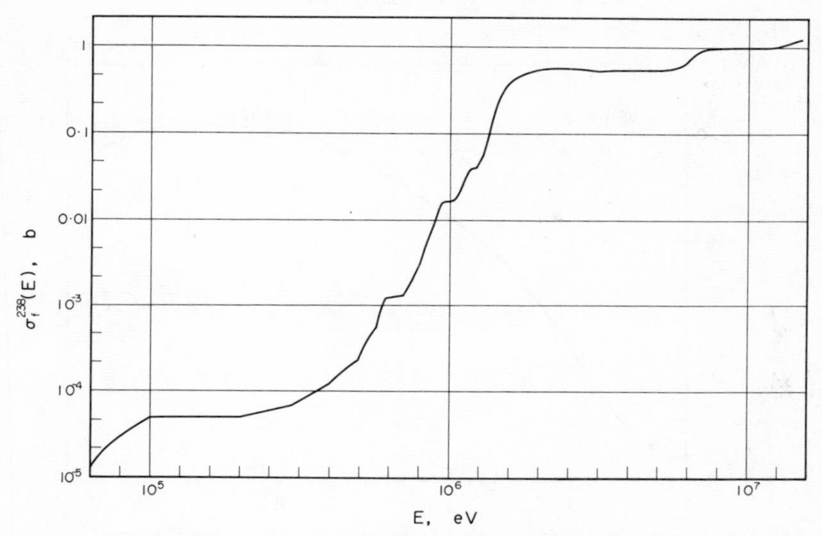

FIG. 1.8. Fission cross-section as a function of energy for U^{238}. Based on ENDF/B data.[4]

Per each fission a certain number of fertile nuclei present in the reactor undergo fast fission. Neutron balance is modified by a fast fission factor customarily denoted by ϵ.

In a fast reactor the fission neutrons are slowed down by inelastic scattering in fertile material, coolant and core structures. Elastic scattering plays an unimportant role. The extent of the slowing down is illustrated in Fig. 1.9 where a (flux) spectrum of a typical fast core is seen. Apart from the neutrons leaking from the fuelled zone of the reactor, energetic neutrons are lost by absorption in the fuel and by radiative capture. Recall that below a certain characteristic threshold absorption by fertile material is solely capture.

The lower end of the energy spectrum in a fast reactor adjoins the resonance energy region from about 10 keV downwards to 1 eV. In a thermal reactor most of the absorptions by fertile material occur in resonances. In order to alleviate the parasitic loss of neutrons, moderating material is introduced there. The moderator accelerates the slowing down process by introducing a viable mechanism of elastic scattering for moderation.

Consider a neutron with an initial energy E_i undergoing an elastic

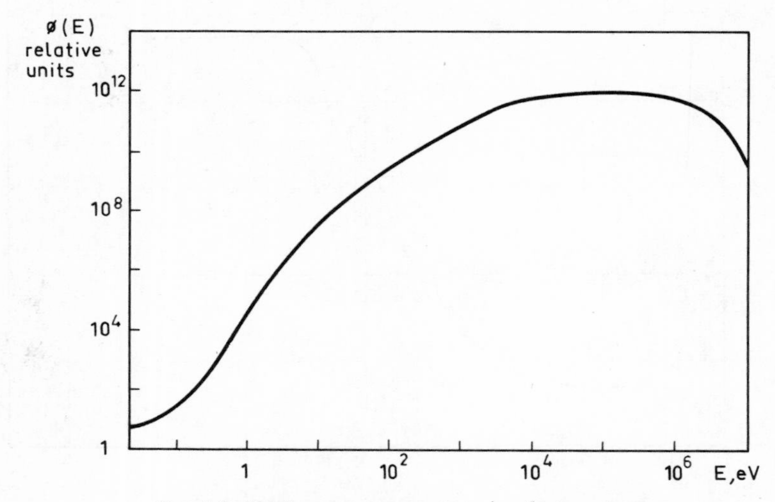

FIG. 1.9. Neutron energy spectrum in a fast reactor.

collision with a nucleus whose atomic mass is A. Let the neutron be scattered through an angle θ in the centre-of-mass system. On the basis of energy and momentum conservation the final energy E_f after the collision is by simple means, e.g. ref. 7, inferred to be

$$E_f = \tfrac{1}{2} E_i [(1+\alpha) + (1-\alpha)\cos\theta] \tag{1.15}$$

where the auxiliary parameter α is defined by

$$\alpha = (A-1)^2/(A+1)^2. \tag{1.16}$$

The maximum energy loss occurs when $\theta = \pi$ and it is equal to

$$\Delta E_{max} = (1-\alpha) E_i. \tag{1.17}$$

In view of eq. (1.17) it is obvious that the lighter the mass A the greater the fraction of the initial neutron energy that can be transferred over into kinetic energy of the target. This is why light materials such as hydrogen, deuterium or graphite are employed as moderators and why light substances are not utilized in fast reactors.

Descending through a transient region below 1 eV, the neutrons enter the stage at which their energies correspond to the energy carried in thermal motions of moderator atoms and molecules. The thermodynamic quasi-equilibrium is characteristic to thermal reactors where the neutron energy spectrum has a major peak between 0.02 and 0.1 eV. At thermal energies the scattering of neutrons incorporates various vibrational and rotational motions of the target. Models governing these phenomena will be discussed in the quantitative chapters of this book. It is important, however, to realize that neutrons may now gain as well as lose energy. An overall energy behaviour of neutrons in a thermal reactor is drawn in Fig. 1.10. The dependent variable in Fig. 1.10 is chosen to be $E\phi(E)$ in order to straighten the graph. This is implicitly the case in many practical situations where one introduces the lethargy variable u as

$$u = \ln \frac{E_0}{E} \tag{1.18}$$

where E_0 denotes an arbitrary energy point. The corresponding spectral form of neutron flux $\phi(u)$ is derived by simply setting the differential quantities $\phi(E)\,dE$ and $\phi(u)\,du$ to be equal. One obtains

$$\phi(u) = E\phi(E). \tag{1.19}$$

The maximum at about 1 MeV in Fig. 1.10 is due to the neutrons released in fissions. The pronounced low energy maximum of

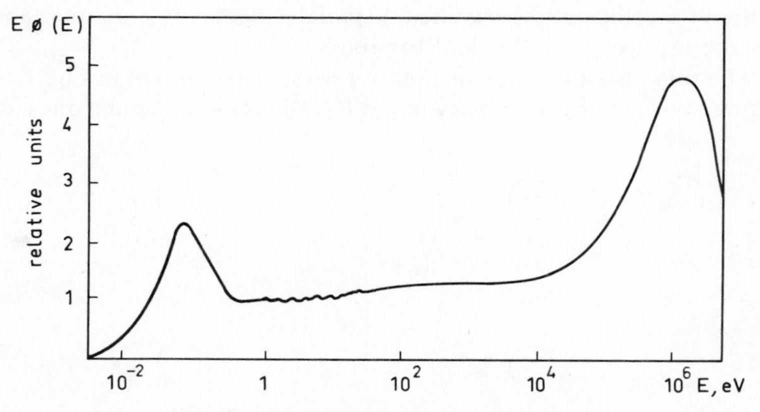

FIG. 1.10. Neutron lethargy spectrum in a thermal reactor.

thermal neutrons can in many cases be closely approximated by the Maxwellian distribution

$$\phi_M(E) = \frac{E}{(kT)^2} \exp(-E/kT) \tag{1.20}$$

with the normalization

$$\int_0^\infty \phi_M(E)\, dE = 1. \tag{1.21}$$

In eq. (1.20) k is Boltzmann's constant and T is the temperature of the medium. Note in particular that the temperature is a parameter of primary importance. For example, the maximum of $\phi_M(E)$ is obtained at $E = kT$. At room temperatures the maximum occurs at an energy tantamount to the neutron speed of 2200 m/s which is commonly taken as a reference point. In the energy scale this corresponds to 0.0253 eV.

1.8. Unit Cell

The spatial heterogeneity of a reactor core arising from separate fuel, coolant and possible moderator regions is to be considered parallel to energy variations of neutrons. Due to the nature of the neutron behaviour and associated characteristic distances of motion without a collision, the significant spatial aspects are most tractable on the level of an individual fuel pin.

The two possible types of fuel rod pattern are shown in Fig. 1.11. Square lattices are customary in LWRs, whereas hexagonal ones are

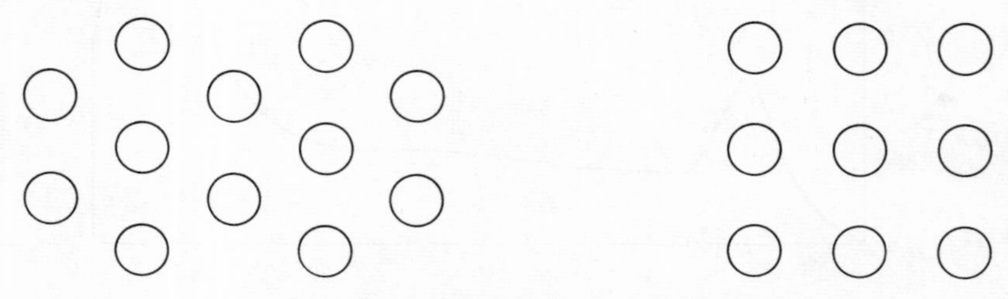

FIG. 1.11. Hexagonal and square fuel configurations.

employed in fast reactors and in some thermal designs as well. From the core physics point of view, the difference is meaningless. It has pure computational relevance in the way of stronger coupling between the neighbouring rods in the hexagonal lattice.

Neglecting the description of the lattice periphery, Fig. 1.11 can be considered as a part of an infinite system. Then all the information is contained in a unit cell consisting of a single fuel rod within a square or hexagon whose exterior is occupied by the coolant and the moderator. The two types of unit cell are shown in Fig. 1.12.

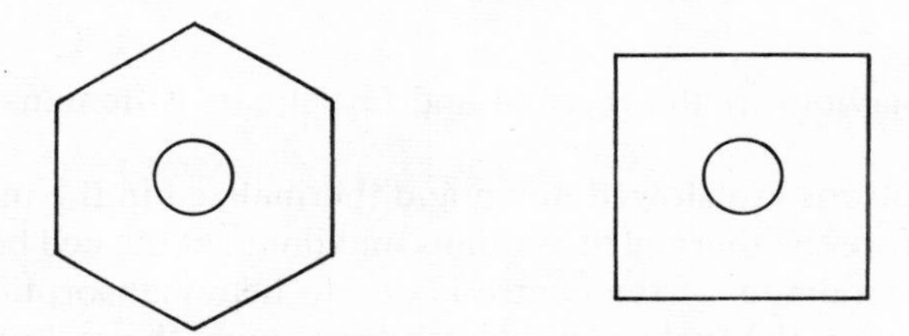

FIG. 1.12. Hexagonal and square unit cells.

In order to combine the energy and spatial variation of the neutron distribution in a unit cell, consider a one-dimensional projection along a symmetry axis of the cells in Fig. 1.12. Taking a LWR as an example and referring to Fig. 1.10, the energy scale is split into three parts: thermal region extending to about 0.6 eV, resonance region from 0.6 eV to 5.5 keV, and fast region above 5.5 keV. The thermal, resonance and fast neutron fluxes ϕ_{th}, ϕ_r and ϕ_f are now defined by

$$\phi_{th}(\mathbf{r}) = \int_0^{E_{th}} \phi(\mathbf{r}, E)\, dE, \tag{1.22}$$

$$\phi_r(\mathbf{r}) = \int_{E_{th}}^{E_r} \phi(\mathbf{r}, E)\, dE \tag{1.23}$$

and

$$\phi_f(\mathbf{r}) = \int_{E_r}^{\infty} \phi(\mathbf{r}, E)\, dE \tag{1.24}$$

where $\phi(\mathbf{r}, E)$ is the neutron flux within the unit cell and $E_{th} = 0.6$ eV, $E_r = 5.5$ keV.

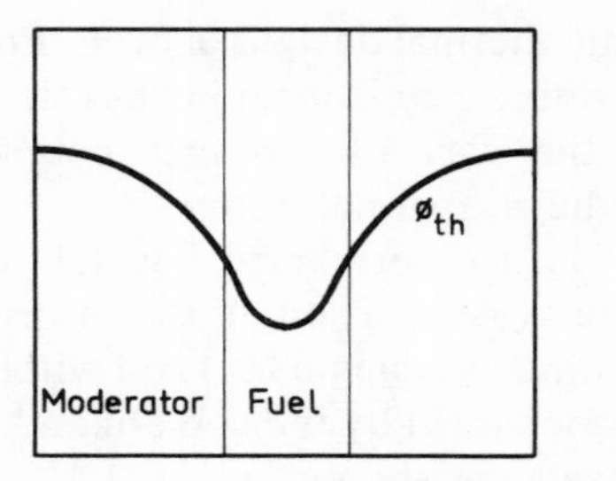

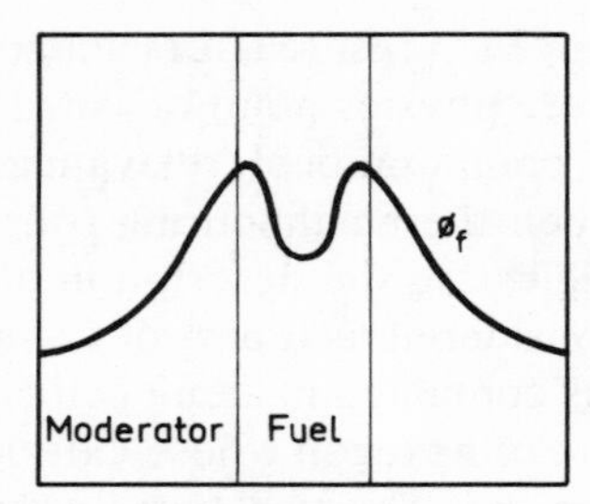

FIG. 1.13. Thermal and fast fluxes in a unit cell.

The behaviour of the thermal and fast fluxes is demonstrated in Fig. 1.13.

The neutrons are slowed down and thermalized in the moderator and therefore the thermal flux attains maximum at the cell boundary, while the minimum at the centre is due to heavy absorption in the outer region of the fuel rod which tends to shield the rest of the fuel region. Some temperature-dependent resonance effects contribute to the depression. Between the two extremes the flux varies monotonically.

As the fast neutrons are released after absorption of thermal neutrons, the fast spectrum also has a local minimum at the centre of the lattice cell. Due to the slowing down there is another local minimum at the edge. The fast flux is increased towards the rod surface and the maximum occurs right inside the rod.

As far as the resonance component of the flux is concerned, it ought to be analysed more quantitatively in order to gain the corresponding insight. At the energies around resonance peaks of heavy absorption the flux $\phi(r, E)$ almost vanishes in the fuel. The flux behaviour in a realistic case is given in Fig. 1.14 where computed results for a mixed U–Pu system are given.[8]

The slight but noticeable increase of the neutron density exhibited in the moderator and cladding is explained by a relatively weak scattering contribution imbedded in the absorption peak of this particular resonance.

To describe the depression of the absorption rate caused by the behaviour of the thermal flux as seen in Fig. 1.13 one usually introduces a quantity called the thermal utilization and denoted by f. It is defined as the ratio of thermal absorption rate in the fuel to

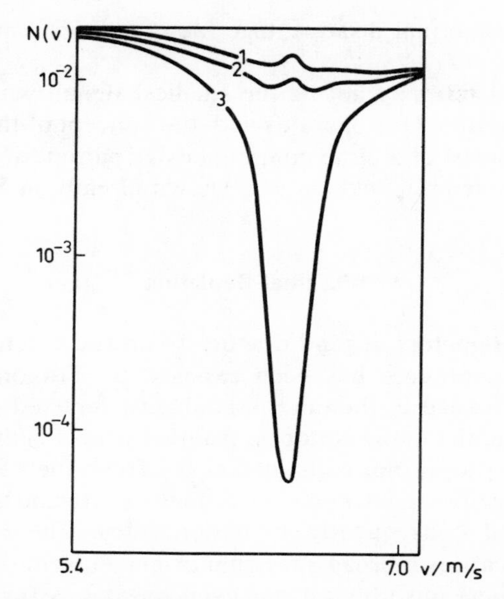

FIG. 1.14. Neutron density distribution in the vicinity of the 1 eV resonance of Pu^{240} as calculated in moderator (1), cladding (2) and fuel (3). Reproduced from ref. 8 with permission of the Technical Research Centre of Finland.

thermal absorption rate in the entire cell. Accordingly

$$f = \frac{\int_{V_F} \Sigma_a^F \phi_{th}(\mathbf{r})\, d\mathbf{r}}{\int_{V_F} \Sigma_a^F \phi_{th}(\mathbf{r})\, d\mathbf{r} + \int_{V_M} \Sigma_a^M \phi_{th}(\mathbf{r})\, d\mathbf{r}} \tag{1.25}$$

where V_F and V_M are the fuel and moderator volumes, respectively. The cladding has been omitted here and therefore $V_F + V_M$ represents the volume of the unit cell.

The practical utility of the thermal utilization concept is provided by the fact that the absorption cross-sections Σ_a^F and Σ_a^M are or can be approximated as constants over the respective regions and the ratio

$$\zeta = \frac{V_F \int_{V_M} \phi_{th}(\mathbf{r})\, d\mathbf{r}}{V_M \int_{V_F} \phi_{th}(\mathbf{r})\, d\mathbf{r}}, \tag{1.26}$$

known as the thermal disadvantage factor, can be computed separately.

A single-pin cell may not be the smallest region where the core analysis is initiated. One operates with the concept of the lattice cell which may consist of a more comprehensive pattern of fuel rods up to an entire assembly such as was shown already in Fig. 1.1.

1.9. Fuel Depletion

Reactor parameters depend drastically on the extent of neutron radiation that the core has been exposed to. For one thing, the primary fuel loaded in the core is gradually depleted of the abundance of the initial fissile isotopes. Another aspect is the accumulation of new isotopes not contained in the fresh fuel. Furthermore, both these factors contribute to changes in the neutron energy spectrum, and consequently in fission rates. The new isotopic composition carries a broad spectrum of nuclei, some of which are fissile and others absorbing if not even poisonous for the neutron economy.

The degree of depletion is measured in terms of burnup which is defined as the amount of energy released per unit mass of fuel. Fuel is now defined to comprehend all fissionable isotopes in the fuel charge excluding other chemical compounds of the fuel materials. Hence the proper unit, for example, in LWRs without Pu recycle is the megawatt day per ton of U. This unit is in common use, whereas Wd/gU might sound less inflated and GJ/kgU would comply with the SI standards.

Suppose that F is the amount of fuel charged in a given reactor volume V. At any later time t the average burnup τ in V is obtained from

$$\tau(t) = \frac{1}{F} \sum_m e_m \int_0^t \int_V \int_0^\infty \Sigma_f^m(\mathbf{r}, E, t')\phi(\mathbf{r}, E, t')\, dE\, d\mathbf{r}\, dt' \tag{1.27}$$

where the sum is extended over all fissionable isotopes with corresponding energy release per fission being denoted by e_m. The values of e_m for certain fissile isotopes appear already in Table 1.2

and can be expressed in watts by means of the equivalence $1 \text{ eV} = 1.6 \times 10^{-19}$ Ws. Accordingly, the numerator of eq. (1.27) represents the reactor power

$$P = \sum_m e_m \int_V \int_0^\infty \Sigma_f^m(\mathbf{r}, E)\phi(\mathbf{r}, E)\, dE\, d\mathbf{r}. \tag{1.28}$$

Neither the burnup τ nor the power P is directly concerned with the fuel consumption where the parasitic absorptions take their toll. These quantities are given in terms of the thermal power (Wt) without the reduction caused by the limited efficiency of conversion to electricity.

The temporal behaviour of the isotopic composition depends on the fuel design. In case the fuel charge is of natural uranium, U^{238} is abundantly present and the Pu^{239} buildup is very substantial. On the other hand, if the fuel is highly enriched, Pu^{239} accumulation and consumption are of lesser importance.

Figure 1.15 shows the relative absorption rates of different

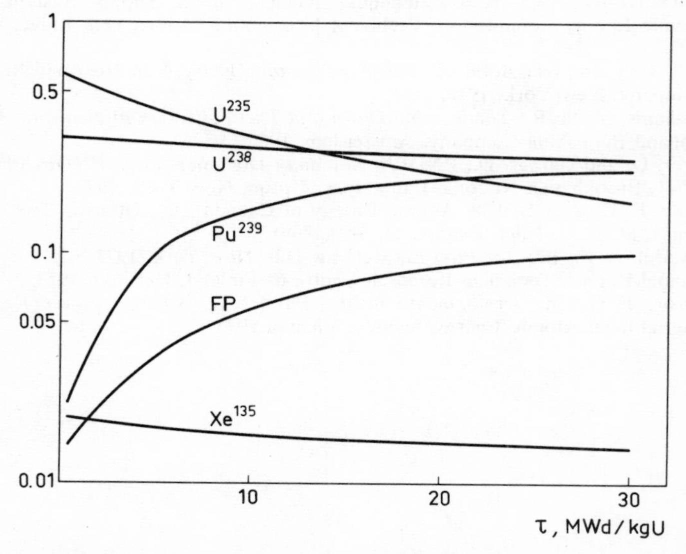

FIG. 1.15. Instantaneous absorption fractions as a function of fuel burnup in a LWR lattice.

isotopes as a function of burnup. The computation is performed for a PWR lattice cell.

In addition to the steady increase of plutonium, the most obvious feature in Fig. 1.15 is the rapid and considerable portion of absorptions due to the fission products. Note that the graphs of the most important fission products Xe^{135} are drawn separately. It is worth while to observe that even within this burnup range up to 25 GWd/tU for the fuel with the initial enrichment of 3.6% the absorption rate of Pu^{239} is greater than that of U^{235}. The fission-to-capture ratio would, however, still result in favour of U^{235}.

By no means is Fig. 1.15 implying that Pu^{239} concentration would increase indefinitely. For uranium-fuelled systems where very high burnups are conceivable it is evident[9] that the absolute amount of Pu^{239} or even Pu^{241} begins to decrease at a certain point.

References

1. DOCKET-RESARA-16, Westinghouse Nuclear Steam Supply System, U.S. Atomic Energy Commission, Technical Information Center, Oak Ridge, Tenn., 1973.
2. Bell, G. I. and Glasstone, S., *Nuclear Reactor Theory*, Van Nostrand Reinhold Company, New York, 1970.
3. Williams, M. M. R., *The Slowing Down and Thermalization of Neutrons*, North-Holland Publishing Company, Amsterdam, 1966.
4. Ozer, O. and Garber, D., ENDF/B Summary Documentation. ENDF-201, BNL 17541, Brookhaven National Laboratory, Upton, New York, 1973.
5. Lane, F. E., AECL-3038, Atomic Energy of Canada Ltd., Ottawa, 1969.
6. James, M. F., *J. Nucl. Energy*, **23**, 517 (1969).
7. Zweifel, P. F., *Reactor Physics*, McGraw-Hill, New York, 1973.
8. Annual Report, Technical Research Centre of Finland, Helsinki, 1973.
9. Tyror, J. G., in *Developments in the Physics of Nuclear Power Reactors*, International Atomic Energy Agency, Vienna, 1973.

CHAPTER 2

Neutron Diffusion

A QUANTITATIVE analysis of the core performance, the distribution of power generation and burnup is based on the study of the angular neutron density maintained in the reactor. All the relevant interactions, neutron generation and leakage are described or approximated by mathematically formulated relations which are assembled to a single equation known as the neutron transport equation.

In view of the subsequent discussion on core analysis, the transport equation and a number of approximate variants pertaining to it are introduced in this chapter. For a more comprehensive derivation and theoretical treatments the reader may wish to consult refs. 1 and 2.

As a matter of fact, only a minor portion of the core management analyses will be based directly on the transport equation, but rather the diffusion approximation shall be employed. However, the intrinsic nature of the diffusion approximation is best understood and justified through its relation to the original formulation and it is therefore most appropriate to introduce these two equations consecutively.

2.1. The Stationary Neutron Transport Equation

While the neutrons traverse in the reactor and interact with the core medium they obey the natural law of conservation. The transport equation is a balance equation expressing the conservation of neutrons in an arbitrary infinitesimal element of the space-velocity phase space. If the system is maintained in a stationary state the rate of neutrons entering the element is equal to the rate of their outflow from the element.

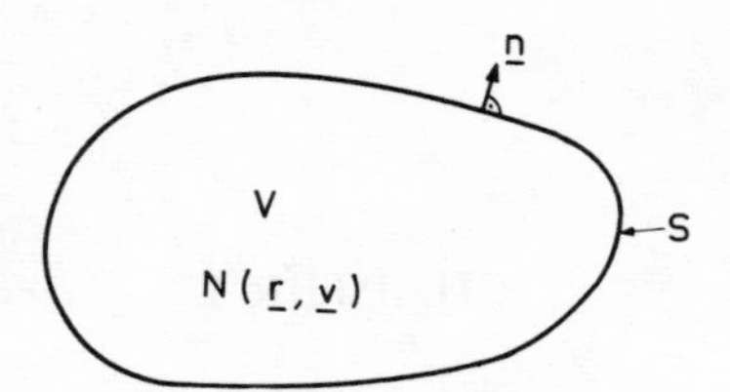

FIG. 2.1. Core volume element.

Neutrons may leave the element due to spatial streaming or different interactions and they enter as source or fission neutrons or after being scattered in. Consider first the streaming of neutrons which is related to the volume element V. Let again $N(\mathbf{r}, \mathbf{v})$ be the neutron density distribution within the volume element V bound by the surface S. Pursuant to the remarks made in section 1.3, $\mathbf{n} \cdot \mathbf{v}\, N\, dS$ is the rate at which neutrons intersect the surface element dS whose outward normal is denoted by $\mathbf{n}$. The rate of crossing the surface is given per unit element in the velocity space. The net leakage across S is given by $\int_S \mathbf{n} \cdot \mathbf{v} N(\mathbf{r}, \mathbf{v})\, dS$. The application of Gauss's divergence theorem yields for the integral

$$\begin{aligned}\int_S \mathbf{n} \cdot \mathbf{v} N(\mathbf{r}, \mathbf{v})\, dS &= \int_V \nabla \cdot \mathbf{v} N(\mathbf{r}, \mathbf{v})\, d\mathbf{r} \\ &= \int_V \mathbf{v} \cdot \nabla N(\mathbf{r}, \mathbf{v})\, d\mathbf{v}\end{aligned} \tag{2.1}$$

where the second equality is just an identity following from the spatial nature of the divergence operator. Incidentally, the quantity $\mathbf{v}N$ appearing in eq. (2.1) is known as the angular current.

According to eq. (1.10), the total interaction rate per unit element in the velocity space is given by $\int_V v\, \Sigma_t N\, d\mathbf{r}$.

Among the positive contributions to the neutron balance consider first scattering. The rate at which in-scattering occurs is readily given by $\int_V \int_\omega v' \Sigma_S(\mathbf{r}, \mathbf{v}') f(\mathbf{r}, \mathbf{v}', \mathbf{v}) N(\mathbf{r}, \mathbf{v}')\, d\mathbf{v}'\, d\mathbf{r}$ where ω denotes the velocity space. The corresponding rate of the inflow of fission neutrons is equal to

$$\chi(v) \int_V \int_\omega v' \nu(v') \Sigma_f(\mathbf{r}, \mathbf{v}') N(\mathbf{r}, \mathbf{v}')\, d\mathbf{v}'\, d\mathbf{r}.$$

In writing down the balance equation, i.e. in assembling the different terms together, one is reminded of the arbitrariness of the volume V which implies that the balance will be fulfilled pointwise. This results in the stationary neutron transport equation

$$\begin{aligned}\mathbf{v}\cdot\nabla N(\mathbf{r},\mathbf{v}) + v\Sigma_t(\mathbf{r},\mathbf{v})N(\mathbf{r},\mathbf{v}) \\ = \int_\omega v'\Sigma_s(\mathbf{r},\mathbf{v}')f_s(\mathbf{r},\mathbf{v}',\mathbf{v})N(\mathbf{r},\mathbf{v}')\,d\mathbf{v}' \\ + \chi(v)\int_\omega v'\nu(v')\Sigma_f(\mathbf{r},\mathbf{v}')N(\mathbf{r},\mathbf{v}')\,d\mathbf{v}' + Q(\mathbf{r},\mathbf{v})\end{aligned} \tag{2.2}$$

where Q refers to the distribution of possible extraneous sources. For the sake of brevity, introduce the notation B for the transport operator

$$\begin{aligned}(B\Phi)(\mathbf{r},\mathbf{v}) = \mathbf{\Omega}\cdot\nabla\Phi(\mathbf{r},\mathbf{v}) + \Sigma_t(\mathbf{r},\mathbf{v})\Phi(\mathbf{r},\mathbf{v}) \\ - \int_\omega \Sigma_S(\mathbf{r},\mathbf{v}')f_S(\mathbf{r},\mathbf{v}',\mathbf{v})\Phi(\mathbf{r},\mathbf{v}')\,d\mathbf{v}'.\end{aligned} \tag{2.3}$$

The angular flux $\Phi(\mathbf{r},\mathbf{v})$ (cf. section 1.2) has been taken as the dependent variable in eq. (2.3). There will be two different kinds of approaches to eq. (2.2). Firstly, there is an inhomogeneous problem where the fission term either vanishes or is fictitiously included in the source. Equation (2.2) reduces to

$$(B\Phi)(\mathbf{r},\mathbf{v}) = Q(\mathbf{r},\mathbf{v}). \tag{2.4}$$

Given the source Q, it can be shown[1] that the solution of eq. (2.4) is uniquely determined as soon as the incident neutron distribution at the surface of the reactor volume is specified. Normally there are no incoming neutrons and the incident distribution is equal to zero.

The problems of the second type with no extraneous sources pertain to an operating reactor. Defining the fission operator F by

$$(F\Phi)(\mathbf{r},\mathbf{v}) = \chi(v)\int_\omega \nu(v')\Sigma_f(\mathbf{r},\mathbf{v}')\Phi(\mathbf{r},\mathbf{v}')\,d\mathbf{v}' \tag{2.5}$$

eq. (2.2) reduces now to an eigenvalue problem of the form

$$(B\Phi)(\mathbf{r},\mathbf{v}) = \lambda(F\Phi)(\mathbf{r},\mathbf{v}) \tag{2.6}$$

where λ represents the eigenvalue. Neutron transport theory predicts that there exists a real non-degenerate smallest eigenvalue and an associated non-negative eigenfunction Φ_0. For $\lambda_0 = 1$ it is seen from eq. (2.6) that this corresponds physically to a selfsustaining situation. The inverse $k = 1/\lambda_0$ is known as the effective multiplication factor of the reactor.

The dominant value k of the inverse eigenvalues determines the criticality of the reactor. If k is equal to unity, the system is critical. $k > 1$ and $k < 1$ correspond to a supercritical and subcritical system, respectively.

A conventional approach to the eigenvalue problem posed in eq. (2.6) is the source iteration scheme where the fission and part of the scattering are converted into an iterative source. The formulation of the problem is that of eq. (2.4).

2.2. Boundary Conditions

Equation (2.2) describes the behaviour of the neutron distribution in an accurate manner unnecessarily detailed for most core fuel management applications. However, the physical boundary conditions always originate from this formulation and are therefore most readily accessible within the same framework.

The neutron angular density remains bounded within the reactor system and vanishes at infinity, i.e.

$$\lim_{|\mathbf{r}|\to\infty} N(\mathbf{r}, \mathbf{v}) \to 0 \tag{2.7}$$

with the exception of special idealizations where one separately assumes a source at infinity. Across interior material discontinuities $N(\mathbf{r}, \mathbf{v})$ is continuous. Let $\mathbf{n}$ denote the normal of such an interior surface S. With respect to an arbitrary direction described by a unit vector $\mathbf{e}$, the continuity relation may then be expressed as

$$\lim_{\epsilon\to 0} N(\mathbf{r} + \epsilon\mathbf{e}, \mathbf{v}) = \lim_{\epsilon\to 0} N(\mathbf{r} - \epsilon\mathbf{e}, \mathbf{v}) \tag{2.8}$$

$$\mathbf{r} \in S \quad \text{and} \quad \mathbf{n}\cdot\mathbf{v} \neq 0.$$

The condition $\mathbf{n}\cdot\mathbf{v} \neq 0$ excludes the neutrons traversing parallel to a plane surface. The density and flux components parallel to a plane in the velocity space but in different media are clearly uncoupled and do not bear the continuity implication.

In order to describe the leakage of neutrons from a convex finite or semi-infinite system one introduces a non-re-entrant surface where there are no neutrons in the inward direction, i.e.

$$N(\mathbf{r}, \mathbf{v}) = 0, \quad \mathbf{v} \cdot \mathbf{n} < 0 \qquad \mathbf{r} \in S \tag{2.9}$$

where again $\mathbf{n}$ is the outward normal. Most often eq. (2.9) is referred to as the vacuum boundary condition. The surrounding medium is simply non-moderating, e.g. purely absorbing or vacuum.

Equation (2.9) is not actually concerned with the speed but the unit direction of the velocity and the relation is rewritten as

$$N(\mathbf{r}, \mathbf{v}) = 0, \quad \mathbf{n} \cdot \mathbf{\Omega} < 0 \qquad \mathbf{r} \in S. \tag{2.10}$$

The geometry of the problem encountered may frequently be reduced by extracting all the symmetry relations pertaining to the given configuration. The entire geometrical structure is obtained by reflection across the symmetry axes involved. Let S now be a surface across which symmetry prevails. For any direction specified by the unit vector $\mathbf{\Omega}$ there is the symmetrical direction such that $(\mathbf{\Omega} - \mathbf{\Omega}') \times \mathbf{n} = \mathbf{0}$ where $\times$ denotes the vector product. The reflection symmetry is then expressed as

$$N(\mathbf{r}, \mathbf{v}) = N(\mathbf{r}, \mathbf{v}') \qquad \mathbf{r} \in S, \quad |\mathbf{v}| = |\mathbf{v}'|, \quad (\mathbf{\Omega} - \mathbf{\Omega}') \times \mathbf{n} = \mathbf{0}. \tag{2.11}$$

The most obvious uses of the reflection symmetry relation are the reductions of the unit cell and the entire core into the smallest distinct pattern.

If the unit cell of an infinite lattice is not symmetrical, for some reason the rotational invariance known as the periodic boundary condition is achieved by the requirement that corresponding to the distribution of emerging neutrons across a boundary there will be inserted an identical distribution of incoming neutrons at the opposite, i.e. periodically equivalent, boundary of the unit cell. The more common reflection condition of eq. (2.11) is a special case of the periodic one. This is illustrated in Fig. 2.2 where a one-dimensional

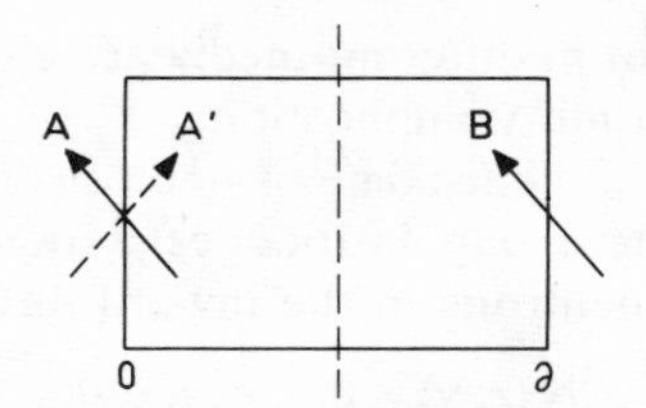

FIG. 2.2. Periodic and reflective boundaries in plane geometry.

periodic system is described by a unit cell. An outflow of neutrons from the cell at $x = 0$ is described by the arrow A. The periodic boundary condition implies there to be an inflow of neutrons at $x = a$, which represents the opposite surface. The arrow B corresponds to the periodicity requirement. If the unit cell, however, possesses a symmetry axis at $x = a/2$, the arrow B is equivalent to the arrow A' and A-A' implements the reflective boundary condition.

Consider finally a problem arising in the substitution of a real lattice or unit cell by an approximate cylindrical one in accordance with Fig. 2.3. The introduction of the cylindrical equivalent cell[3] reduces the number of spatial dimensions and the periodic condition along the hexagonal or square boundary cannot be strictly fulfilled. The reflection condition can mathematically be imposed, but it tends to distort the real behaviour by yielding too high fluxes in the moderator.[2, 3] One circumvents the problem satisfactorily by placing an isotropically scattering ring around the equivalent cell. This procedure is known as the white boundary condition and it will be discussed again subsequently.

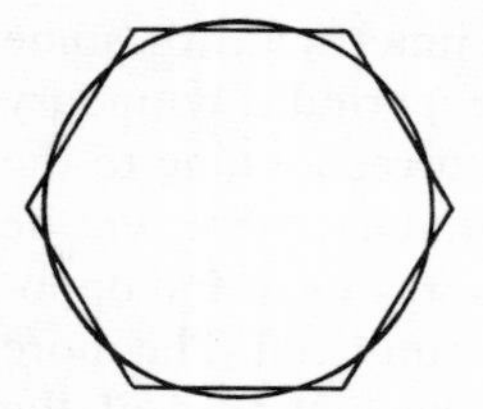

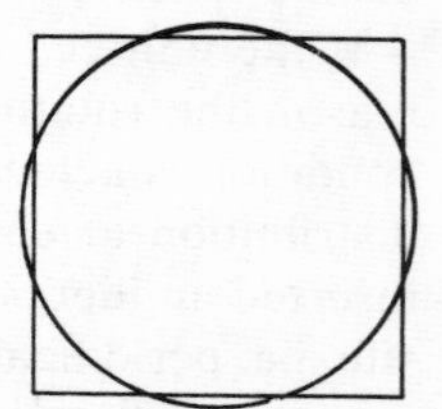

FIG. 2.3. Equivalent cylindrical cells.

2.3. P_1 and Diffusion Approximations

Among the variety of approximations applied to eq. (2.2) there is a certain class in which the angular direction variable $\mathbf{\Omega}$ is eliminated from the transport equation. Loosely speaking, this category can be referred to as the diffusion approximation, because it usually yields for the total flux $\phi(\mathbf{r}, v)$ an equation similar to the equations governing proper diffusion processes.

The rationale of commencing the reduction of independent variables from $\mathbf{\Omega}$ also resides in the physical implications of diffusion. To a large extent neutrons behave in an isotropic manner independently of $\mathbf{\Omega}$. The fission neutrons are born isotropically and scattering at large produces only weak preference of directions. The anisotropy arises mainly from the streaming of neutrons in a heterogeneous configuration and propagates within a distance of a few mean free paths $(\Sigma_t)^{-1}$.

It is also pointed out that to a good approximation the angular scattering distribution function f_s appearing in the transport equation depends individually not on the directions $\mathbf{\Omega}$ and $\mathbf{\Omega}'$ but, being rotationally invariant, on the scattering angle whose cosine is given by $\mu_0 = \mathbf{\Omega} \cdot \mathbf{\Omega}'$.

The separation of $\mathbf{\Omega}$ from the velocity necessitates the notation $\Phi(r, E)$ for the angular flux. Consequently, the cross-sections and the angular scattering distribution function $f_s(r, E', E, \mu_0)$ are viewed as functions of energy and direction separately. To the lowest non-trivial order $\Phi(r, E, \mathbf{\Omega})$ is approximated in $\mathbf{\Omega}$ by the two-term expression

$$\Phi(\mathbf{r}, E, \mathbf{\Omega}) = \frac{1}{4\pi}[\phi(\mathbf{r}, E) + 3\mathbf{\Omega} \cdot \mathbf{J}(\mathbf{r}, E)] \tag{2.12}$$

where $\phi(\mathbf{r}, E)$ is consistently the total flux

$$\phi(\mathbf{r}, E) = \int_{4\pi} \Phi(\mathbf{r}, E, \mathbf{\Omega})\, d\mathbf{\Omega} \tag{2.13}$$

and $\mathbf{J}(\mathbf{r}, E)$, called the neutron current, is the integral of the angular current

$$\mathbf{J}(\mathbf{r}, E) = \int_{4\pi} \mathbf{\Omega}\Phi(\mathbf{r}, E, \mathbf{\Omega})\, d\mathbf{\Omega} = \int_{4\pi} v\mathbf{\Omega}N(\mathbf{r}, E, \mathbf{\Omega})\, d\mathbf{\Omega}. \tag{2.14}$$

In order to derive the diffusion approximation, the transport equation

$$\mathbf{\Omega} \cdot \nabla\Phi(\mathbf{r}, E, \mathbf{\Omega}) + \Sigma_t(\mathbf{r}, E)\Phi(\mathbf{r}, E, \mathbf{\Omega}) = \int\int \Sigma_s(\mathbf{r}, E')f_s(\mathbf{r}, E', E, \mu_0)\Phi(\mathbf{r}, E', \mathbf{\Omega}')\, dE'\, d\mathbf{\Omega}' + Q(\mathbf{r}, E, \mathbf{\Omega}) \tag{2.15}$$

is first integrated over all $\mathbf{\Omega}$. The fission term in eq. (2.2) is already implied to be isotropic and it will not be carried within the derivation.

Prior to performing the integrations, consider the substitution of eq. (2.12) in the scattering term of eq. (2.15). The angular scattering distribution f_s is expanded in terms of spherical harmonics

$$f_s(\mathbf{r}, E, E', \mu_0) = \sum_{l=0}^{\infty} \frac{2l+1}{4\pi} f_{sl}(\mathbf{r}, E, E')P_l(\mu_0). \tag{2.16}$$

In view of the trial function (2.12) the orthogonality of Legendre polynomials causes the contribution to vanish from the terms higher than $l = 0, 1$. The scattering term reduces to

$$\int\int \Sigma_s f_s \Phi\, dE'\, d\mathbf{\Omega}' = \frac{1}{4\pi}\int_0^{\infty} \Sigma_s(\mathbf{r}, E')[f_{s0}(\mathbf{r}, E', E)\phi(\mathbf{r}, E') + 3f_{s1}(\mathbf{r}, E', E)\mathbf{\Omega} \cdot \mathbf{J}(\mathbf{r}, E')\, dE']. \tag{2.17}$$

Inserting Φ from eq. (2.12) on the left-hand side of eq. (2.15) and performing the integrations, one arrives at

$$\nabla \cdot \mathbf{J}(\mathbf{r}, E) + \Sigma_t(\mathbf{r}, E)\phi(\mathbf{r}, E) = \int_0^{\infty} \Sigma_s(\mathbf{r}, E')f_{s0}(\mathbf{r}, E', E)\phi(\mathbf{r}, E')\, dE' + Q_0(\mathbf{r}, E) \tag{2.18}$$

with

$$Q_0(\mathbf{r}, E) = \int_{4\pi} Q(\mathbf{r}, E, \mathbf{\Omega})\, d\mathbf{\Omega}. \tag{2.19}$$

The simple formulae to perform the integrations are not reproduced here, but the reader may find them in ref. 2, for example.

Consistent with the two unknowns allowed in eq. (2.12) and consequently in (2.18), another equation is needed to supplement eq. (2.18). Multiplying eq. (2.15) by $\mathbf{\Omega}$, inserting (2.17) and integrating over all $\mathbf{\Omega}$ yields

$$\frac{1}{3}\nabla\phi(\mathbf{r}, E) + \Sigma_t(\mathbf{r}, E)\mathbf{J}(\mathbf{r}, E) = \int_0^\infty \Sigma_s(\mathbf{r}, E')f_{s1}(\mathbf{r}, E', E)\mathbf{J}(\mathbf{r}, E')\, dE' + \mathbf{Q}_1(\mathbf{r}, E) \qquad (2.20)$$

where

$$\mathbf{Q}_1(\mathbf{r}, E) = \int_{4\pi} \mathbf{\Omega} Q(\mathbf{r}, E, \mathbf{\Omega})\, d\mathbf{\Omega}. \qquad (2.21)$$

Equations (2.18) and (2.20) constitute the so-called P_1 approximation. This is because eq. (2.12) is viewed as including the two first terms P_0 and P_1 of a Legendre expansion of Φ. Due to the fact that no separability of variables is implied so far, eqs. (2.18) and (2.20) couple the flux and current through the scattering kernels in addition to the coupling exhibited on the left-hand sides of the equations. As such, eq. (2.18) is obtained rigorously, whereas a true expansion of the angular flux in a series of higher Legendre polynomials would provide a further term in the streaming part of eq. (2.20) (cf. refs. 2, 4 and 5).

At this point it should be emphasized that the P_1 approximation is customarily employed in two distinct applications. In the slowing down calculations, where energy or lethargy is the independent variable of major consideration, great concern is devoted to eq. (2.20) in order to obtain a relation between the flux and the current. This relation is stated in the form

$$\mathbf{J}(\mathbf{r}, E) = -D(r, E)\nabla\phi(\mathbf{r}, E) \qquad (2.22)$$

where the diffusion coefficient $D(\mathbf{r}, E)$ is obtained by an approximation of eq. (2.20), viz. it is assumed that the slowing down intensity is relatively smooth and in the integrand $\Sigma_s(E')f_{s1}(E', E)\mathbf{J}(E')$ the energies E and E' can be interchanged. By means of this argument which is far from rigorous and observing that the integral

$$\int_0^\infty f_{s1}(\mathbf{r}, E', E)\, dE' = \int_0^\infty \int \mu_0 f_s(\mathbf{r}, E', E, \mu_0)\, d\mu_0\, dE' = \bar{\mu}_0(\mathbf{r}, E) \qquad (2.23)$$

yields the average value $\bar{\mu}_0(\mathbf{r}, E)$ of the cosine of the scattering angle at which neutrons deviate after a collision at E', one obtains for $D(\mathbf{r}, E)$ from eq. (2.20)

$$D(\mathbf{r}, E) = \tfrac{1}{3}[\Sigma_t(\mathbf{r}, E) - \bar{\mu}_0(\mathbf{r}, E)\Sigma_s(\mathbf{r}, E)]^{-1}. \quad (2.24)$$

In deriving eq. (2.24) it is also tacitly assumed that the source Q is isotropic and $\mathbf{Q}_j$ vanishes in eq. (2.20). The combination $\Sigma_t - \bar{\mu}_0\Sigma_s$ appearing in eq. (2.24) is called the transport cross-section Σ_{tr}.

The other uses of the P_1 approximation in situations where the space variable is of main interest involve less onerous concern of the diffusion coefficient. The energy dependence is there handled by a multigroup approach which will be discussed in the next section. The proper transport equation is first reduced to a form independent of energy and hence there will be no energy coupling on the left-hand side of the expression corresponding to the $\Sigma_s f_{s1}\mathbf{J}$ term in (2.20).

Diffusion theory is based on the P_1 approximation essentially by the introduction of the diffusion coefficient in eq. (2.22) which is substituted in eq. (2.18). The zeroth Legendre coefficient f_{s0} in the expansion (2.16) has a simple physical meaning attached to it, viz.

$$f_{s0}(\mathbf{r}, E', E) = \int_{-1}^{1} f_s(\mathbf{r}, E', E)\, d\mu_0 \quad (2.25)$$

is the probability at $\mathbf{r}$ that a scattering event at E' will transfer the neutron to the energy interval dE at E. The product $\Sigma_s f_{s0}$ is defined as the scattering kernel

$$\Sigma_s(\mathbf{r}, E', E) = \Sigma_s(\mathbf{r}, E') f_{s0}(\mathbf{r}, E', E). \quad (2.26)$$

Inserting eqs. (2.22) and (2.26) in eq. (2.18) one obtains the energy-dependent diffusion equation

$$(B\phi)(\mathbf{r}, E) = Q_0(\mathbf{r}, E) \quad (2.27)$$

or

$$-\nabla \cdot D(\mathbf{r}, E)\nabla\phi(\mathbf{r}, E) + \Sigma_t(\mathbf{r}, E)\phi(\mathbf{r}, E) - \int_0^\infty \Sigma_s(\mathbf{r}, E', E)\phi(\mathbf{r}, E')\, dE' = Q_0(\mathbf{r}, E).$$

The continuity requirements and boundary conditions imposed on the angular flux $\Phi(r, E, \mathbf{\Omega})$ in the previous section are inherent to the flux and current. The continuity of Φ across internal interfaces implies the continuity of ϕ and $\mathbf{J}$ as well. Furthermore, the conditions at the exterior or symmetry boundaries S are to be derived consistently with the P_1 approximation itself. All these conditions can be generally expressed by

$$a(\mathbf{r}, E)\phi(\mathbf{r}, E) + b(\mathbf{r}, E)\mathbf{n}\cdot\nabla\phi(\mathbf{r}, E) = 0 \qquad e \in S. \tag{2.28}$$

Some further discussion on the form of eq. (2.28) in practical considerations will be contained in subsequent sections.

Equation (2.27) provides the underlying basis for a large number of practical algorithms employed in the core analyses where the space and energy variation of the neutron distribution is to be solved. From the derivation it is obvious that the streaming term is crucial to the diffusion approximation where the angular variables are not displayed explicitly. In core designs there are numerous situations where neutrons diffuse anisotropically and the streaming has to be considered more carefully. An interesting example is met in a gas-cooled fast reactor core[6] where axial voidage occurs between fuel rods, necessitating the use of two diffusion coefficients, one accounting for axial and the other for radial streaming.

Diffusion theory is generally speaking valid in systems where the interaction of neutrons with the medium results in an asymptotic mode of the neutron behaviour. The asymptotic state of neutrons is one where correlation diminishes between the existing distribution and the distribution neutrons obey while injected into the system. Let c denote the average number of secondary neutrons per collision,

$$c = \frac{\Sigma_s + \nu\Sigma_f}{\Sigma_t}. \tag{2.29}$$

In general the diffusion theory approximation is a good one when $|c - 1| \ll 1$ (cf. the discussion in ref. 1). This condition is adequately closely fulfilled in a nuclear reactor while it manifestly fails to prevail in a purely absorbing system or, at the other extreme, in a highly multiplicative system such as a nuclear explosion.

2.4. Multigroup Methods

Regardless whether the transport or the diffusion equation is concerned, the energy variable is most often treated by a discretization into a finite number of energy intervals, points or modes. The explicit execution of the reduction may vary substantially according to the particular computational requirements, the unifying feature being that eqs. (2.15) and (2.27) are replaced by a system of similiar equations each of which is energy-independent. The two approaches described in this section are the common multigroup method and the method of modal approximations.

Consider first the neutron transport equation (2.15) where the notation of scattering kernel $\Sigma_s(\mathbf{r}, E', E, \mu_0) = \Sigma_s(r, E)f_s$ is introduced and Q corresponds to either an extraneous source or the fission source

$$Q = \frac{\lambda}{4\pi}\chi(E)\int_0^\infty \nu(E')\Sigma_f(\mathbf{r}, E')\Phi(\mathbf{r}, E', \mathbf{\Omega})\,dE'. \tag{2.30}$$

In modal approximations the angular flux Φ is expanded in terms of energy trial functions $\alpha_i(E)$

$$\Phi(\mathbf{r}, E, \mathbf{\Omega}) = \sum_{i=1}^{N} \alpha_i(E)\psi_i(\mathbf{r}, \mathbf{\Omega}). \tag{2.31}$$

The set of trial functions $\{\alpha_i\}$ is determined *a priori*. The expression in eq. (2.31) is inserted in eq. (2.15) which is concurrently multiplied by the weighting functions $\beta_i(E)$, consecutively for $i = 1,\ldots, N$. The resulting equations are integrated over all E.

While it is quite nonessential but shortens only the notation orthogonality

$$\int_0^\infty \alpha_i(E)\beta_j(E)\,dE = \delta_{ij} \tag{2.32}$$

is assumed here. Defining the $N \times N$ matrices

$$[\Sigma_t]_{ij} = \int_0^\infty \Sigma_t(\mathbf{r}, E)\alpha_i(E)\beta_j(E)\,dE, \tag{2.33}$$

$$[\Sigma_s]_{ij} = \int_0^\infty dE\beta_j(E)\int_0^\infty \Sigma_s(\mathbf{r}, E', E, \mu_0)\alpha_i(E')\,dE' \tag{2.34}$$

and the N-component vectors

$$[\mathbf{Q}]_i = \int_0^\infty Q(\mathbf{r}, E, \mathbf{\Omega})\beta_i(E)\, dE', \tag{2.35}$$

$$[\boldsymbol{\psi}]_i = \psi_i \tag{2.36}$$

the transport equation is reduced to the form

$$\begin{aligned}\mathbf{\Omega} \cdot \nabla\boldsymbol{\psi}(\mathbf{r}, \mathbf{\Omega}) + \Sigma_t(\mathbf{r})\boldsymbol{\psi}(\mathbf{r}, \mathbf{\Omega})\\ = \int_{4\pi} \Sigma_s(\mathbf{r}, \boldsymbol{\mu}_0)\boldsymbol{\psi}(\mathbf{r}, \mathbf{\Omega}')\, d\mathbf{\Omega}' + \boldsymbol{Q}(\mathbf{r}, \mathbf{\Omega}).\end{aligned} \tag{2.37}$$

The notation $\mathbf{\Omega} \cdot \nabla\boldsymbol{\psi}$ is intended to imply that the gradient operator is valid for each of the $\boldsymbol{\psi}_n$'s separately. The matrix Σ_t is subsequently assumed to be diagonal, which can be brought about by a simple similarity transformation. In the eigenvalue problem one specifies the source vector $\mathbf{\Omega}$ as

$$\boldsymbol{Q}(\mathbf{r}, \mathbf{\Omega}) = \lambda\,\Sigma_f(\mathbf{r}) \int_{4\pi} \boldsymbol{\psi}(\mathbf{r}, \mathbf{\Omega})\, d\mathbf{\Omega} \tag{2.38}$$

where the fission cross-section matrix Σ_f is given by

$$[\Sigma_f]_{ij} = \frac{1}{4\pi}\int_0^\infty dE\beta_j(E)\chi(E)\int_0^\infty \nu(E')\alpha_i(E')\, dE'. \tag{2.39}$$

The physical meaning attached to the elements of the cross-section matrices $[\Sigma]_{ij}$ corresponds to the particle transfer taking place from the ith to the jth mode. The aptitude of the formalism for providing a useful energy approximation is dependent on the number of modes made available and on the particular selection of the trial and weighting functions. The multigroup method used most commonly is tantamount to assuming that the weighting functions are all equal to unity and the trial functions are composed of unit step functions by splitting the energy variable into a finite number of intervals (Fig. 2.4).

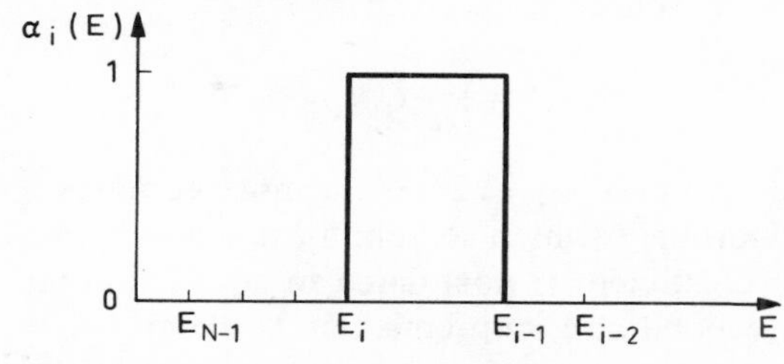

FIG. 2.4. Base energy function in the multigroup method.

$\alpha_i(E)$ is now defined to vanish outside $\Delta E_i = (E_i, E_{i-1})$ and to be equal to unity within this interval. In the N-group method $E_N = 0$ and E_0 is chosen to be equal to the largest value of neutron energy in the system, typically of the order of 10 MeV.

In order to derive the P_1 multigroup equations, eqs. (2.18) and (2.20) are integrated over the energy groups ΔE_i, $i = 1, 2, \ldots, N$. Group fluxes $\phi_i(\mathbf{r})$ and currents $\mathbf{J}_i(\mathbf{r})$ which correspond to the angular flux modes $\psi_i(r)$ in eq. (2.31) are defined by

$$\phi_i(\mathbf{r}) = \int_{\Delta E_i} \phi(\mathbf{r}, E)\, dE \tag{2.40}$$

and

$$\mathbf{J}_i(\mathbf{r}) = \int_{\Delta E_i} \mathbf{J}(\mathbf{r}, E)\, dE \tag{2.41}$$

respectively.

The total cross-section of the ith group is defined by

$$\Sigma_{ti}(\mathbf{r}) = \int_{\Delta E_i} \frac{\Sigma_t(\mathbf{r}, E)\phi(\mathbf{r}, E)\, dE}{\phi_i(\mathbf{r})}. \tag{2.42}$$

The energy coupling in eq. (2.18) due to the scattering kernel is treated by introducing a transfer matrix Σ_s whose elements $[\Sigma_s]_{ij}$ will be given in the form

$$[\Sigma_s]_{ij} = \int_{\Delta E_j} dE \int_{\Delta E_i} \frac{\Sigma_s(\mathbf{r}, E', E)\phi(\mathbf{r}, E')\, dE'}{\phi_i(\mathbf{r})}. \tag{2.43}$$

By these definitions eq. (2.18) has the discretized form

$$\nabla \cdot \mathbf{J}_i(\mathbf{r}) + \Sigma_{ti}(\mathbf{r})\phi_i(\mathbf{r}) = \sum_{j=1}^{N} \Sigma_{sji}(\mathbf{r})\phi_j(\mathbf{r}) + q_i(\mathbf{r}) \tag{2.44}$$

where the group source q_i is given as

$$q_i(\mathbf{r}) = \int_{\Delta E_i} Q_0(\mathbf{r}, E)\, dE. \tag{2.45}$$

The discretization of eq. (2.20) is far more complex if one desires to proceed rigorously, which seldom turns out to be practical. Once the diffusion coefficient is postulated in eq. (2.22) one may simply define a corresponding group constant D_i from

$$\mathbf{J}_i(\mathbf{r}) = -D_i(\mathbf{r})\nabla\phi_i(\mathbf{r}). \tag{2.46}$$

To obtain a less heuristic relation eq. (2.20) is written in the form where the transport cross-section is inserted. The pertinent relation in the ith group then has the form

$$\tfrac{1}{3}\nabla\phi_i(\mathbf{r})+\Sigma_{\text{tr}i}(\mathbf{r})\mathbf{J}_i(\mathbf{r})=0. \tag{2.47}$$

It is readily observed that Σ_{tr} is to be averaged over the current

$$\Sigma_{\text{tr}i}(\mathbf{r})\mathbf{J}_i(\mathbf{r})=\int_{\Delta E_i}[\Sigma_t(\mathbf{r},E)-\bar{\mu}_0(\mathbf{r},E)\Sigma_s(\mathbf{r},E)]\mathbf{J}(\mathbf{r},E)\,dE \tag{2.48}$$

while all the other group constants are flux weighted averages over the energy group concerned. The diffusion coefficient will be given by

$$D_i(\mathbf{r})=\frac{1}{3\Sigma_{\text{tr}i}(\mathbf{r})}. \tag{2.49}$$

Equation (2.47) actually allows Σ_{tr} to be a matrix rather than a scalar, and this implies an additional assumption hidden in eq. (2.49).

In a proper P_1 multigroup scheme eq. (2.20) warrants further sophistication since $D(\mathbf{r},E)$ really is an operator and not an ordinary function. The set of the multigroup equations would encompass a diffusion coefficient matrix with non-zero off-diagonal elements D_{ij}, whereas only diagonal elements D_i will be considered in all subsequent uses in this text.

At this point, if not earlier, one is obliged to pose a question about the extent to which the crude approximations made above distort the behaviour of the neutron distribution when calculated employing the multigroup approach. The reason why the treatment is in common use is to be found in the extensive experience with operating reactors and numerous experimental comparisons. These have achieved a fairly standard practice in terms of the number of groups applied in any particular situation. Furthermore, those energy regions where the cross-sections vary rapidly, e.g. the resonance region, are treated differently and will be discussed separately.

The P_1 approximation in the direction variables can be derived from eq. (2.37) where the transport equation is already discretized in energy. The derivation is analogous to one given in section 2.3 and will not be reproduced here, but the reader may find it in ref. 2. Since there will be no energy coupling in the transfer kernels of eqs. (2.18) and (2.20), this approach will avoid the explicit difficulty of obtaining the group diffusion coefficient. From the practical point of view, however, the group averages are to be taken over angular fluxes and currents whose approximation may turn out to be difficult.

The dilemma in the multigroup methods boils down to the fact that the group constants appearing in the equation are averages over the fluxes, i.e. over the solution to this equation. The problem is iterative in nature. In a single problem the iteration is alleviated by the use of previously accumulated flux distributions. For another thing, if the multigroup equation is primarily aimed to describe the space dependence the detailed energy spectrum is less essential.

To conclude this section certain modifications are made in eq. (2.44). The scattering within the ith group is combined on the left-hand side by defining the macroscopic total removal cross-section

$$\Sigma_{ri} = \Sigma_{ti}(\mathbf{r}) - \Sigma_{sii}(\mathbf{r}). \tag{2.50}$$

The source term is also rewritten to include the fission source. In the standard notation effective fission cross-section $\nu\Sigma_{fi}$ in the energy group i is defined by

$$\nu\Sigma_{fi}(\mathbf{r}) = \int_{\Delta E_i} \frac{\nu(E)\Sigma_f(\mathbf{r}, E)\phi(\mathbf{r}, E)\, dE}{\phi_i(\mathbf{r})} \tag{2.51}$$

and the multigroup diffusion equation is then expressed as

$$-\nabla \cdot D_i(\mathbf{r})\nabla\phi_i(\mathbf{r}) + \Sigma_{ri}(\mathbf{r})\phi_i(\mathbf{r}) = \sum_{\substack{j=1 \\ j \neq i}}^{N} \Sigma_{sji}(\mathbf{r})\phi_j(\mathbf{r}) + \lambda\chi_i \sum_{j=1}^{N} \nu\Sigma_{fj}(\mathbf{r}) \tag{2.52}$$

where χ_i represents the fission spectrum yield in the ith group, i.e.

$$\chi_i = \int_{\Delta E_i} \chi(E)\, dE. \tag{2.53}$$

The multigroup method outlined above is not commonly used to condense resonance cross-section information where some auxiliary techniques have been devised. Resonance effects will be discussed in section 4.2 and further in section 7.3.

2.5. Approximate Boundary Conditions

Both the energy discretization and the P_1 approximation require the boundary conditions originally imposed on the angular neutron density and flux to be modified in a manner consistent with the approximations themselves. This is simply because the solution of the resulting diffusion equation would otherwise be undefined.

As far as the multigroup technique is concerned, the energy variable appears implicitly and the group parameters are defined as to preserve reaction rates. Consequently, all the spatial or angular conditions in the continuous treatment can be formulated for each given group without further concern.

The group fluxes and the normal components of the group currents are continuous across any interior material discontinuities occurring in the system. It is mainly the exterior or symmetry boundaries which necessitate further consideration. Reiterating the one-group non-re-entrant condition at the surface S whose outward normal is denoted by $\mathbf{n}$

$$\Phi_i(\mathbf{r}, \mathbf{\Omega}) = 0 \qquad \mathbf{r} \in S, \quad \mathbf{n} \cdot \mathbf{\Omega} < 0 \tag{2.54}$$

one substitutes this by an integral relation known as the Marshak boundary condition. In the P_1 approximation the procedure is equivalent to requiring the incoming current to vanish at the free surface, i.e.

$$\int_{n \cdot \Omega < 0} \mathbf{n} \cdot \mathbf{\Omega} \Phi_i(\mathbf{r}, \mathbf{\Omega})\, d\mathbf{\Omega} = 0 \qquad \mathbf{r} \in S. \tag{2.55}$$

Upon insertion of the P_1 expression

$$\Phi_i(\mathbf{r}, \mathbf{\Omega}) = \frac{1}{4\pi}(\phi_i(\mathbf{r}) + 3\mathbf{\Omega} \cdot \mathbf{J}_i(\mathbf{r})) \tag{2.56}$$

in eq. (2.55) one arrives at

$$\tfrac{1}{2}\phi_i(\mathbf{r}) - \mathbf{n} \cdot \mathbf{J}_i(\mathbf{r}) = \tfrac{1}{2}\phi_i(\mathbf{r}) + D_i(\mathbf{r})\mathbf{n} \cdot \nabla\phi_i(\mathbf{r}) = 0 \qquad \mathbf{r} \in S. \tag{2.57}$$

The condition in eq. (2.57) is obtained under a physically realistic assumption on the current. To prove that eq. (2.57) is consistent with the P_1 procedure one has to resort to more theoretical arguments. It can be shown[7] that if the P_1 approximation were derived by means of variational analysis eq. (2.57) would follow concomitantly with the diffusion approximation.

Suppose that the flux $\phi_i(\mathbf{r})$ would be extrapolated linearly at the boundary. In plane geometry, for example $\phi_i(x) = \phi_i(x_s) + \phi'_i(x_s)(x - x_s)$, it is seen that the extrapolated flux vanishes at $\phi_i(r_s)/\mathbf{n} \cdot \nabla\phi_i(r_s)$, which quantity is known as the linear extrapolation distance and is found to be equal to $d_i = 2D_i$ or $2/3\Sigma_{tr}$ of length. The linear extrapolation distance is not to be confused with the proper extrapolation length or extrapolated endpoint which is obtained from an asymptotic mode of the proper transport solution.[1] This quantity would be equal to $0.71045\ 1/\Sigma_{ti}$.

The utility of this boundary condition is not restricted to free surfaces, but it may be used at interior boundaries comprising non-moderating regions as well, provided that the non-multiplicativity extends over a few mean free paths. This would correspond to rather massive control rods inserted in the core. Another example is found at the core–reflector interfaces. The core is surrounded by the reflector in order to make some part of the leaked neutrons return.

To illustrate the flux behaviour in the vicinity of the interface, consider Fig. 2.5 where the thermal and fast fluxes are shown at a LWR core–reflector boundary. The thermal and fast fluxes are defined in accordance with the discussion in sections 1.7 and 1.8. It is seen that a remarkable peak of the thermal flux occurs in the water

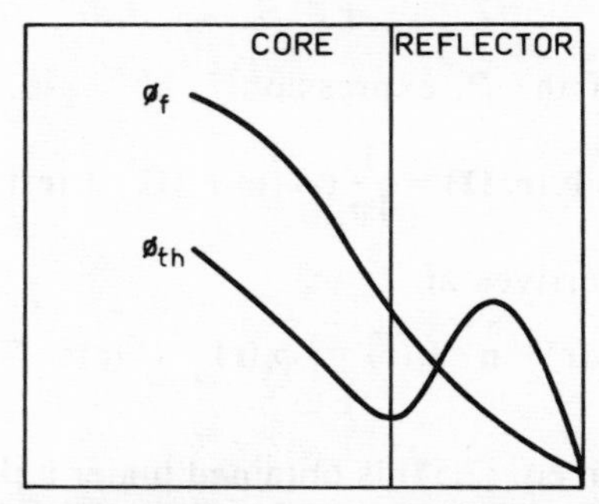

FIG. 2.5. Thermal and fast fluxes at the LWR core–reflector boundary.

reflector due to fast neutrons slowing down there. The computational routines available may on occasions be inadequate or too costly to run in order to reproduce the peak in calculations. One may desire to solve the diffusion equation within the core only and to apply eq. (2.57) at the interface. It is most obvious that the extrapolation distances for these two groups are vastly different and, indeed, even of opposite sign. In practice, the group extrapolation lengths should be computed separately and using a more accurate method than is employed in the core diffusion analysis.

Generalizing the boundary condition in eq. (2.57) to read

$$a_i(\mathbf{r})\phi_i(\mathbf{r}) + b_i(\mathbf{r})\mathbf{n} \cdot \nabla\phi_i(\mathbf{r}) = 0 \qquad \mathbf{r} \in S \tag{2.58}$$

one has essentially all the relevant conditions in one. Clearly, the choice

$$\frac{b_i}{a_i} = d_i = 2D_i \tag{2.59}$$

would correspond to the Marshak boundary condition. If the reactor system is enclosed by a reflector whose thickness is several mean free paths, the exterior boundary is very closely approximated by the condition at infinity, i.e.

$$\phi_i(\mathbf{r}) = 0, \quad \mathbf{r} \in S \tag{2.60}$$

which is equivalent to $b_i = 0$, $a_i \neq 0$ in eq. (2.58).

Across an axis of reflection symmetry the normal component of the current vanishes corresponding to $a_i = 0$, $b_i \neq 0$ in eq. (2.58). Finally, in the periodical boundary condition the emerging current at the surface point $\mathbf{r}_s$ should be equal to the entering current at the surface located at $\mathbf{r}_s + n\boldsymbol{\tau}$ where the unit vector represents the periodicity shift.

In the preceding discussion it referred to an approach where solutions are obtained by means of imposing boundary conditions arrived at by using a more accurate method. In this category there is a popular technique known as the albedo method.[8] The inward and outward group currents J_i^{in} and J_i^{out} are given at the surface by

$$J_i^{\text{in}} = \int_{\mathbf{n}\cdot\boldsymbol{\Omega}<0} |\mathbf{n} \cdot \boldsymbol{\Omega}|\Phi_i(\mathbf{r}, \boldsymbol{\Omega})\, d\boldsymbol{\Omega} = \frac{1}{2}[\phi_i(\mathbf{r}) - 2\mathbf{n} \cdot \mathbf{J}_i(\mathbf{r})] \tag{2.61}$$

and

$$J_i^{\text{out}} = \int_{\mathbf{n}\cdot\mathbf{\Omega}>0} \mathbf{n}\cdot\mathbf{\Omega}\Phi_i(\mathbf{r},\mathbf{\Omega})\,d\mathbf{\Omega} = \frac{1}{2}[\phi_i(\mathbf{r}) + 2\mathbf{n}\cdot\mathbf{J}_i(\mathbf{r})] \qquad (2.62)$$

respectively. Letting $\mathbf{J}^{\text{in}}$ denote the vector whose components are J_i^{in} and similarly $\mathbf{J}^{\text{out}}$ the vector with components J_i^{out} one defines the albedo matrix $\boldsymbol{\alpha}$ by

$$\mathbf{J}^{\text{in}} = \boldsymbol{\alpha}\,\mathbf{J}^{\text{out}}. \qquad (2.63)$$

Note that the superscripts "in" and "out" refer to the core and not to the reflector. For all practical purposes the albedo matrix would be the lower diagonal because the thermal outward current excites no fast current to enter the core. The elements of $\boldsymbol{\alpha}$ are precomputed by an accurate method based upon transport theory, and eq. (2.63) is used as the boundary condition for the diffusion theory solution.

A resembling variant to the albedo method is derived by observing that eqs. (2.61) and (2.62) can be rearranged to yield

$$\phi_i(\mathbf{r}) = J_i^{\text{out}} + J_i^{\text{in}} \qquad (2.64)$$

and

$$\mathbf{n}\cdot\mathbf{J}_i(\mathbf{r}) = \frac{1}{2}(J_i^{\text{out}} - J_i^{\text{in}}). \qquad (2.65)$$

The boundary condition in eq. (2.63) can then be replaced by an equivalent statement connecting the surface flux to the current[9]

$$\boldsymbol{\phi} = \gamma\mathbf{J} \qquad (2.66)$$

where $\boldsymbol{\phi} = (\phi_1, \phi_2, \ldots, \phi_N)$ and $\mathbf{J} = (\mathbf{n}\cdot J_1, \mathbf{n}\cdot J_2, \ldots, \mathbf{n}\cdot J_N)^T$. This system of linear relations is now used at the boundary.

2.6. Diffusion in Various Reactor Systems

In section 1.7 the overall energy spectrum was discussed and a qualitative idea was developed on the spectral characteristics in fast and thermal reactors. To pursue further the discussion of the core behaviour it is worth while to have a similar view on the spatial properties of neutron diffusion.

Consider again the group diffusion equation

$$-\nabla \cdot D_i(\mathbf{r})\nabla\phi_i(\mathbf{r}) + \Sigma_i(\mathbf{r})\phi_i(\mathbf{r}) = q_i(\mathbf{r}) \tag{2.67}$$

given in the preceding sections. From the neutronics point of view the neutron distribution $\phi_i(\mathbf{r})$ has to be analysed over the entire core in order to examine to what extent neutrons detect the size and the heterogeneities of the system. The influence of the boundaries and heterogeneities propagates in the core with the detailed correlation being determined by the solution of eq. (2.67).

Regarding the group source q_i as a known function, the solution depends on the two parameters D_i and Σ_i in addition to the boundary conditions. In the most simplified case of a homogeneous infinite medium the flux due to a planar source obeys the exponential function $\exp(-r/L_i)$ where r is the distance from the source and L_i is known as the diffusion length

$$L_i = \sqrt{\frac{D_i}{\Sigma_i}}. \tag{2.68}$$

L_i is well defined even in a heterogeneous medium where it represents a local relaxation length as accurately as the diffusion equation is valid, i.e. where the source is not too close.

The diffusion length may be averaged over the entire system and over all energy groups. By this procedure one obtains a single integral parameter which can be used to classify the neutron behaviour in the system to a crude degree. In Table 2.1 certain data are collected from ref. 10 to facilitate a simple comparison. Only the reactor concepts which employ a single reactor vessel are included in Table 2.1, because the comparison is not directly applicable to the pressure tube reactors such as the natural uranium heavy water reactor CANDU and the steam generating heavy water reactor (SGHWR).

Table 2.1 exhibits a large variation of the diffusion length when the light water or graphite moderated cases are compared. This is due to the superior moderating characteristics graphite has. In fact, heavy water would imply even longer diffusion lengths. Looking at it in another way, the light water cores have to be designed much larger when measured in terms of the diffusion length.

TABLE 2.1
Core and Diffusion Parameters of Certain 3000 MW *Reactors*[10]

Reactor type	Average diffusion length, cm	Core diameter in diffusion lengths	Core volume, m^3
PWR, pressurized watter reactor	1.8	190	40
BWR, boiling water reactor	2.2	180	60
HTGR, high temperature gas-cooled reactor	12.0	65	430
LMFBR, sodium-cooled fast breeder	5.0	35	6
GCFR, gas-colled fast breeder	6.5	33	11

The relatively high power densities in LWRs contribute to the compactness in absolute units of length. Fast breeder reactors are designed to produce extremely high power densities and their size remains small no matter whether measured in absolute units or in diffusion lengths.

The general behaviour of the LWR is typically understood within the framework of two-group analysis. Figures 1.13 and 2.5 demonstrate the basic differences in the fast and thermal groups when regarded at the level either of a pin cell or of the entire core. The thermal flux is sensitive to the temperature of the system. In some thermal reactors, e.g. HTGRs and SGHWRs, the thermal neutrons will experience at least two temperatures. The rethermalization between the two thermal spectra make two thermal groups unavoidable and the total number of groups reaches at least four or even more. Despite the narrow energy range pertinent to the fast reactors, the spectral effects have important consequences there and the global analysis requires the use of about ten groups.

2.7. Integral Form of the Transport Equation

The neutron transport equation (2.15) possesses a variant form of an integral equation particularly applicable to certain occasions, among which the lattice cell calculations are most important. This form of the equation will be derived in this section via brief physical arguments.

Let the angular neutron flux at $\mathbf{r}$ be again denoted by $\Phi(\mathbf{r}, E, \mathbf{\Omega})$. All the neutrons moving in the specified direction $\mathbf{\Omega}$ have undergone the latest collision or are born on the straight line passing $\mathbf{r}$ in the direction $\mathbf{\Omega}$ as depicted in Fig. 2.6.

Consider the contribution to $\Phi(\mathbf{r}, E, \mathbf{\Omega})$ arising from neutrons at a distance s along $\mathbf{\Omega}$. Assuming isotropic scattering, the effective

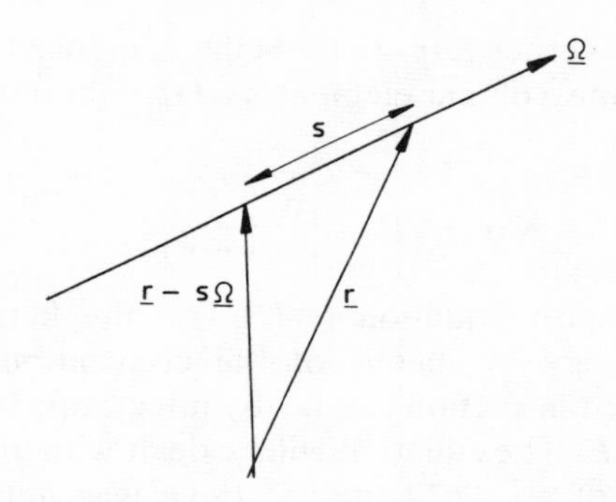

FIG. 2.6. Neutron trajectory and notation for the integral equation.

source combining the inscattering and extraneous sources would be at $\mathbf{r}' = \mathbf{r} - s\mathbf{\Omega}$

$$q(\mathbf{r}', E, \mathbf{\Omega}) = \frac{1}{4\pi}\left[\int_0^\infty \Sigma_s(\mathbf{r}', E', E)\phi(\mathbf{r}', E')\, dE' + Q_0(\mathbf{r}', E)\right] \tag{2.69}$$

where $\Sigma_s(\mathbf{r}, E', E)$ was defined in (2.26). Isotropic scattering implies that $f_s = 1/(4\pi)f_{s0}$ in the scattering kernel of eq. (2.15).

While traversing the distance s, neutrons undergo interactions and the fraction which ever reaches $\mathbf{r}$ is given by $\exp(-\Sigma_t s)q$. Allowing the total cross-section to vary along s, one obtains the angular flux by integrating over all s

$$\Phi(\mathbf{r}, E, \mathbf{\Omega}) = \int_0^\infty \exp(-\alpha(\mathbf{r}, \mathbf{r}', E)q(\mathbf{r} - s\mathbf{\Omega}, E, \mathbf{\Omega})\, ds \tag{2.70}$$

where

$$\alpha(\mathbf{r}, \mathbf{r}', E) = \int_0^s \sum_t(\mathbf{r} - s'\mathbf{\Omega}, E)\, ds' \tag{2.71}$$

is known as the optical distance between the points r and r'. Upon integrating over all $\mathbf{\Omega}$, eq. (2.67) is transformed to express the total flux

$$\phi(\mathbf{r}, E) = \int_V K(\mathbf{r}, \mathbf{r}', E)\left[\int_0^\infty \Sigma_s(\mathbf{r}', E', E)\phi(\mathbf{r}', E')\, dE' + Q_0(\mathbf{r}', E)\right] d\mathbf{r}' \tag{2.72}$$

with the transport kernel $K(\mathbf{r}, \mathbf{r}', E)$ being composed of the optical thickness and of the volume element $ds\, d\mathbf{\Omega}$. Observing that $d\mathbf{r}' = s^2\, ds\, d\mathbf{\Omega}$, one has

$$K(\mathbf{r}, \mathbf{r}', E) = \frac{e^{-\alpha(\mathbf{r}, \mathbf{r}', E)}}{4\pi|\mathbf{r} - \mathbf{r}'|^2}. \tag{2.73}$$

The integral transport equation (2.72) and the kernel (2.73) are discretized in energy by means of the conventional multigroup technique described in section 2.4, i.e. by integrating it over discrete energy intervals ΔE_i. The equation will be dealt with in Chapter 7.

The derivation of eq. (2.72) given above was not directly concerned with the integro-differential equation (2.15). These two can be inferred to be equivalent as far as the determination of the total flux is concerned.[2]

References

1. Case, K. M. and Zweifel, P. F., *Linear Transport Theory*, Addison-Wesley, Reading, Mass., 1967.
2. Bell, G. I. and Glasstone, S., *Nuclear Reactor Theory*, Van Nostrand Reinhold Company, New York, 1970.
3. Honeck, H. C., BNL-5826, Brookhaven National Laboratory, Upton, New York, 1961.
4. Ferziger, H. J. and Zweifel, P. F., *The Theory of Neutron Slowing Down in Nuclear Reactors*, Pergamon Press, Oxford, 1966.
5. Butler, M. K. and Cook, J. M., in *Computing Methods in Reactor Physics*, Gordon & Breach, New York, 1968.
6. Cerbone, R. J., *J. Br. Nucl. Energy Soc.* **12**, 409 (1973).
7. Toivanen, T., *Nucl. Sci. Engng.* **25**, 275 (1966).
8. Delp, D. L. *et al.*, GEAP-4598, General Electric Co., San Jose, Calif., 1964.
9. Pedersen, T. and Kirkegaard, P., M-1607, Danish Atomic Energy Commission, Risö, 1973.
10. Fröhlich, R., in *Numerical Reactor Calculations*, International Atomic Energy Agency, Vienna, 1972.

CHAPTER 3

Core Heat Transfer

HEAT transfer calculations enter the core fuel management analysis for two reasons of indispensable significance. Firstly, the power generation capability of the core is confined by the thermohydraulic parameters which have to remain within the ranges designed. The local temperatures and pressures have an influence on the fuel and core structures and the operating conditions must have certain margins to the imminent phenomena associated with disturbed behaviour of fuel and core materials. Any core management decisions have to be reconciled with the margins. Secondly, regardless of the reactor type concerned, the fuel temperature affects the fuel–neutron interaction and the cross-sections cannot be calculated without knowing the fuel temperature. Furthermore, the coolant tends to moderate neutrons and therefore the state of coolant encompasses a feedback as well. There the temperature and pressure determine the coolant density by which the coolant cross-sections are specified. The importance of the coolant density feedback is magnified in the designs where the coolant simultaneously serves as the moderator.

The treatment in this chapter touches the heat transfer problems and calculational procedures to the depth found appropriate from the fuel management point of view, whereas no attempt will be made to cover the topic in a detailed manner. The questions dealt with are mainly those which arise in light water cooled reactors. This would mean LWRs in the first place, while one should recall that even the SGHWR has heat transfer characteristics quite similar to those of BWR.

3.1. Temperature Profile in Fuel Pin Cell

In general, the rate of power generation varies somewhat across the fuel pin, particularly in thermal reactors where the exterior radial part of the fuel shields the centre, resulting in a depression of the thermal flux as already shown in Fig. 1.13. For the purposes of the present discussion the radial variation causes no qualitative effect on the temperature distribution shown in Fig. 3.1. Figure 3.1 depicts the relative temperature distribution from the centreline of the fuel pin to the centre of the coolant region. To a good approximation one can ignore axial heat conduction, since the radial temperature gradient across the half-thickness $T_f - T_s$ is far more substantial. T_s is the temperature of fuel surface.

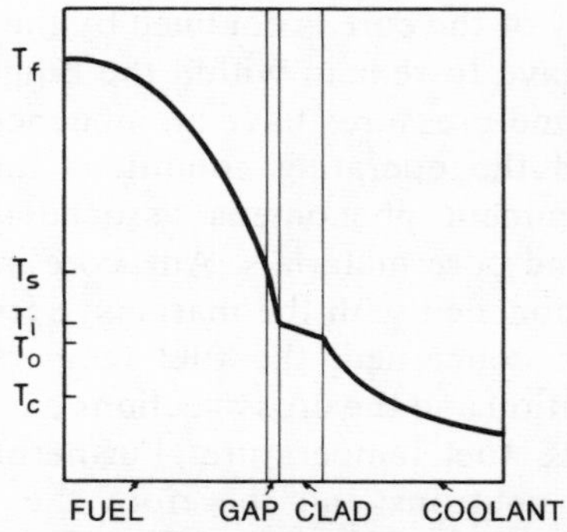

FIG. 3.1. Temperature distribution across a unit cell.

The narrow gap between the fuel pellet and cladding is designed to tolerate the thermal expansion and the swelling of fuel in addition to the volume occupied by fission gases. Relative to fuel and cladding gap conductance is poor, causing the steepest temperature gradient to occur there. Both zircaloy and stainless steel clad possess efficient heat transfer properties and the temperature varies smoothly across the cladding. In the subsequent discussion the inner and outer cladding temperatures are denoted by T_i and T_o, respectively. Particularly in LWRs the maximum cladding temperatures are important and will exhibit the limiting parameters in transient operation.

The heat transfer mechanisms from the cladding to the coolant are strongly dependent on the axial position and the temperature drop from T_o to the average coolant or moderator temperature. T_o varies pursuant to the particular regime of heat transfer.

With proceeding burnup, uranium oxide loses some of its heat conductance due to cracking within the outer pellet region. Simultaneously one has to recall the relaxation of the flux depression. Fission gases are accumulated within the gap and thereby may reduce the conductance. In the regular operation, however, the gap is designed to close, thus improving heat conduction.

In the fast and even some thermal reactors the fuel pellets are annular with a void in the centre. As a result the maximum fuel temperature is reduced, facilitating the operation of the cores where the fuel melting is the primary limiting aspect. Fuel melting should be avoided in the cores of any type because the concomitant volume expansion cannot be withheld. Note that the melting point of UO_2 is about 2800°C.

The heat transfer analysis should be an iterative one, where the causes and consequences from power generation in the fuel to the coolant flow would be taken into account in both directions. In the following a brief summary will be given separately on each step. To avoid the insistence of any specific value of temperature and pressure, an attempt is made to present the material in relative terms while the reader may find the pertinent realistic parameter values for different reactor types in the literature.

Recognizing that each individual phenomenon represents a small facet of the overall heat transfer chain, no equations are derived from the fundamental underlying principles which can be found in the literature.[1,2] It is also important to realize that in contrast to neutron diffusion, reactor heat transfer is based to a large extent on correlations obtained experimentally.

3.2. Heat Conduction in Fuel

Assuming a constant power generation rate in a given volume element of the core, one obtains the corresponding volumetric heat source q_v from eq. (1.28) by averaging over the volume element

$$q_v = \overline{e\Sigma_f\phi}. \tag{3.1}$$

The temperature distribution in the fuel is governed by the ordinary heat conduction equation

$$\nabla \cdot \lambda(\mathbf{r}) \nabla T(\mathbf{r}) + q_v = 0 \tag{3.2}$$

where λ denotes the thermal conductivity of the fuel. Typically, the value of λ for UO_2 falls within the range 2–4 W/m °C, depending on the temperature. Due to the axial conduction being negligible, eq. (3.2) can be solved in cylinder geometry. Upon integrating the one-dimensional equation

$$\frac{1}{r}\frac{d}{dr}\left(r\lambda \frac{dT}{dr}\right) + q_v = 0 \tag{3.3}$$

one obtains first

$$\lambda \frac{dT}{dr} = -\frac{1}{2} q_v r \tag{3.4}$$

where the constant of integration vanishes when the boundary condition $dT/dr = 0$ at $r = 0$ is imposed. Equation (3.4) is then integrated over the fuel pin to yield

$$\Lambda(T, T_s) = \int_{T_s}^{T} \lambda(T)\, dT = \frac{1}{4} q_v (r_s^2 - r^2) \tag{3.5}$$

where r_s refers to radius of the pin and T denotes the temperature at the distance r from the centre. The integrations have been developed in detail in order to introduce the linear heat rating q' defined as the rate of power generation per unit fuel rod length, i.e. $q' = q_v \pi r_s^2$, whereupon the thermal conductivity integral Λ can be expressed as

$$\Lambda(T, T_s) = \frac{q'}{4\pi}\left[1 - \left(\frac{r}{r_s}\right)^2\right]. \tag{3.6}$$

Equation (3.6) combines the two basic quantities which are employed on this occasion. Letting $r = 0$ in eq. (3.6), one obtains a relation between the centre and surface temperatures in the form $\Lambda(T_f, T_s) = q'/4\pi$. q' is sometimes referred to as the linear power density.

Based on knowledge of the thermal conductivity integral, the temperature can be determined at any cylindrical surface within the fuel pellet provided that the temperature is known at an outer surface. The fuel can be divided into a number of radial increments and the solution is obtained by the recursive application of eq. (3.6). One may even let the volumetric heat generation rate q_v vary stepwise and remain constant over each single radial increment only. Thereby the flux depression can be accounted for.

In the vicinity of the nominal conditions of operation, the fuel temperatures are obtained in practice from tabulations computed *a priori*. Even if the integrated thermal conductivity is needed, $\Lambda(T, T_{\text{ref}})$ is usually established for certain reference temperatures T_{ref}. In case the surface temperature is altered, the corresponding integral is interpolated from

$$\Lambda(T, T_s) = \Lambda(T, T_{\text{ref}}) - \Lambda(T_s, T_{\text{ref}}). \tag{3.7}$$

Associated with the volumetric heat source and linear heat rating one often comes across the surface heat flux q'' defined as

$$q'' = \frac{q'}{2\pi r}. \tag{3.8}$$

3.3. Fuel-to-cladding Heat Transfer

The conductance of the gap between the fuel and the clad is rather difficult to analyse. Fortunately, it really is not too essential from the present viewpoint. Heat transfer incorporates here a component of conduction, convection and radiation. The width of the gap is reduced and may even close due to thermal expansion and swelling of the fuel. Mechanical contact between fuel and clad would improve the heat transfer, but poses a problem of whether the interfacial pressure is too high and would cause yielding of the cladding.

In computer simulation[3] all the occurring phenomena are embodied in the gap conductance or total gap heat transfer coefficient α_g to relate the temperature gradient $T_s - T_i$ and the surface heat flux

$$q_s'' = \alpha_g(T_s - T_i). \tag{3.9}$$

Note that q_s'' is known on the basis of the volumetric heat generation rate. Typically, the values of α_g vary around 2500 W/m^2 °C in LWRs corresponding to the temperature difference of about

200°C across the gap. At least in the LMFBR the gap conductance problem is quite similiar and the method of ref. 3 covers in fact both reactor types.

Due to the excellent heat conductivity that the cladding materials possess, heat conduction across the cladding is a relatively simple problem and the temperature gradient ($T_o - T_i$) can be obtained by a direct integration of the heat conduction equation (3.3). Expressing the source term by the linear heat rating one obtains

$$T_o - T_i = \frac{q'}{2\pi\lambda_c} \ln \frac{r_o}{r_i} \tag{3.10}$$

where λ_c denotes the heat conductivity of the cladding and r_i and r_o correspond to the inner and outer radii of the cladding. In practice, these parameters must include the description of the corrosion and oxidation effects in the cladding.

3.4. Axial Temperature Distribution

Heat transfer from cladding to coolant incorporates some extremely difficult problems due to the variation of the transfer mechanisms along the vertical axis of the coolant flow channel. It was indicated earlier that the heat transfer analysis is an iterative calculation where only the radial temperature distribution has been considered so far.

To give a rough estimate of the axial temperature distribution consider an infinitesimal increase dT_c in the coolant temperature arising from the heat generated within the corresponding fuel rod length dz. Letting the flow rate of the coolant be denoted by $G = \rho v$, where ρ and v refer to the coolant density and velocity, respectively, one obtains a balance equation for the enthalpy rise over dz

$$c_p AG \, dT_c = q' \, dz \tag{3.11}$$

with c_p representing the specific heat of the coolant fluid and A being the cross-sectional area of the flow channel. Integrating eq. (3.11) from the channel inlet to z yields

$$T_c(z) = T_c(o) + \int_o^z \frac{q(z')}{AGc_p(z')} dz'. \tag{3.12}$$

The solution of the diffusion equation which determines the axial distribution may be approximated by

$$\phi(z) = \phi_o \sin\left(\pi \frac{z+d}{Z+2d}\right) \tag{3.13}$$

over the core height. d denotes the axial extrapolation distance. Pursuant to eq. (3.13), the flux vanishes at $z = -d$ and $z = Z + d$.

The linear heat rating $q'(z)$ obeys the same form

$$q'(z) = q'_m \sin\left(\pi \frac{z+d}{Z+2d}\right) \tag{3.14}$$

where q'_m denotes the maximum value. Upon substituting the expression (3.14) in eq. (3.12) one obtains

$$T_c(z) = T_c(o) + \frac{q'_m Z}{\pi c_p A G}\left[1 - \cos\frac{\pi z}{Z}\right] \tag{3.15}$$

where c_p is tacitly assumed to be constant along the flow channel. The extrapolation distance d is ignored in eq. (3.15).

To complete the discussion of radial heat transfer which involves the temperature gradient from the outer cladding temperature T_o to the radially averaged coolant temperature T_c, one employs the surface heat flux relation

$$q''_o = \alpha_0(T_o - T_c) \tag{3.16}$$

where α_o is the cladding-to-coolant heat transfer coefficient and q''_o is the surface heat flux at the cladding–coolant interface. q''_o is given by

$$q''_o = \frac{q'}{2\pi r_o} \tag{3.17}$$

[cf. eq. (3.8)]. By virtue of eq. (3.14) it is seen that the temperature difference $(T_o(z) - T_c(z))$ obeys a chopped sinusoidal dependence on z, and $T_o(z)$ is determined by

$$T_o(z) = T_c(z) + \frac{q'_m}{2\pi\alpha_o r_o}\sin\left(\pi\frac{z+d}{Z+2d}\right) \tag{3.18}$$

where T_c was given in eq. (3.15).

The model described above is a simplified one which ignores the gradual change of phase taking place in a realistic case. The axial temperature distributions obtained from eqs. (3.15) and (3.18) are drawn in Fig. 3.2. It is seen in Fig. 3.2 that the maximum clad temperature occurs above the mid-core $z = Z/2$. Given a smooth heat source distribution one would expect the maximum inner cladding and fuel temperatures to occur closer to the mid-plane the closer one approaches to the fuel centerline. Typical LWR values of $T_o - T_c$ range from 10°C in BWR to 20°C in PWR.

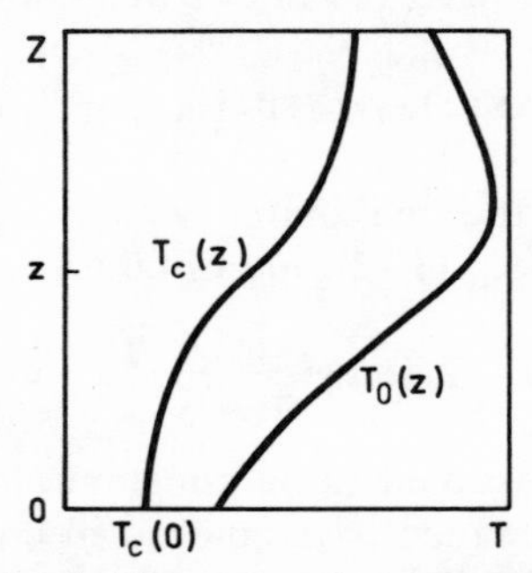

FIG. 3.2. Axial cladding and coolant temperature profiles.

In case the specific heat varies noticeably with z, the flow channel is divided into regions where c_p is assumed to be a constant. Equation (3.12) is then integrated separately over each region and at each step the new c_p values are computed from the pressure distribution. One should also observe the appearance of two phases of the coolant. This problem is considered in the following section.

3.5. Coolant Flow in LWRs

Coolant undergoes a change of phase while passing the LWR core. Two-phase flow phenomena have been thoroughly discussed in refs. 4 and 5, while this section deals with the problem only to the extent required for understanding the coolant feedback background.

The presence of the steam vapour phase in the flow is described in terms of the void fraction defined as the ratio of the vapour volume to the total flow volume. In the flow of the two-phase mixture the vapour component moves at a higher velocity v_v than the liquid velocity v_l. The ratio

$$S = \frac{v_v}{v_l} \tag{3.19}$$

is known as the slip ratio. The void fraction is expressed by

$$\alpha = \frac{1}{1 + S\dfrac{1-x}{x}\dfrac{\rho_v}{\rho_l}} \tag{3.20}$$

where x denotes the ratio of the mass flow rate of the vapour to the mass flow rate of the two-phase mixture. This quantity is known as the flow quality. In eq. (3.20) ρ_v and ρ_l denote the steam and liquid densities, respectively.

The void profile in a flow channel is drawn in Fig. 3.3.

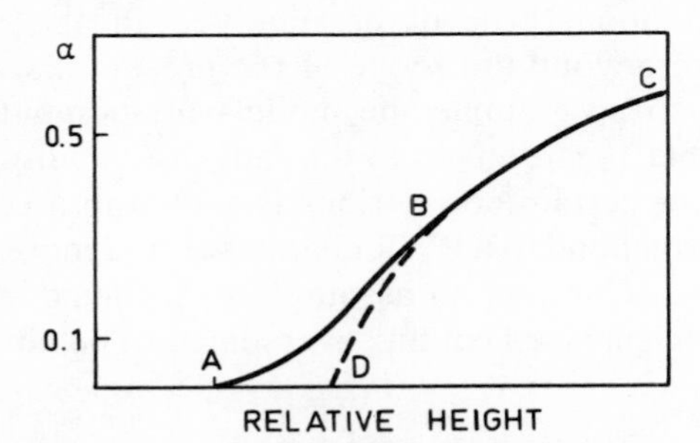

FIG. 3.3. Void fraction in a flow channel.

At the core inlet the flow is in liquid phase. In conservative PWR designs the coolant remains in the single phase and the core outlet is below point A, where the system begins to experience formation of subcooled voidage. In more modern PWRs there is significant boiling in the subcooled region AB, the void fraction reaching a few per cent even in the nominal conditions. The steam bubbles are still recondensing while being transported away from the vicinity of the rod surface. In the hottest channels the temperature exceeds saturation and bulk boiling takes place. Large regions of the core are voidable during transients and over-power operation.

Relatively speaking, subcooled boiling is less important in BWRs which operate over the entire boiling regime depicted in Fig. 3.3. The void fraction reaches typically 50% at the core outlet contributing a major feedback effect on the axial power distribution. Note that the liquid is saturated along the curve *DBC* beyond which bulk boiling occurs.

Void fraction predictions for operating conditions are based upon semiempirical correlations which relate the enthalpy, pressure and other pertinent parameters to void fraction, flow quality or slip ratio. Among the variety of correlations those put forward by Bowring[6] or Rouhani and Axelsson[7] are applicable over the entire range of Fig. 3.3. Within the subcooled regime one can apply the model of Thom as supplemented by Tong[2] or the correlation due to Levy.[8] Sha has presented a generalized correlation[9] involving certain free parameters which effectively facilitate interpolation between some other uniquely determined models.

The computational procedures involved in the void fraction correlations are beyond the scope of the present discussion. It is a simple matter to programme the models for computer and apply them with proper regard given to the range of validity. In Fig. 3.4 a comparison[10] of certain correlations is shown in a case where the parameters correspond to a PWR channel heated more than a typical average channel. Contrary to assumptions intrinsic in some of the correlations, the surface heat flux is assumed to be non-uniform and

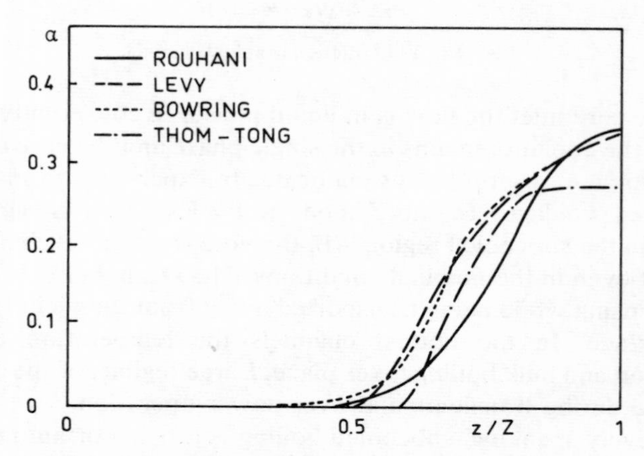

FIG. 3.4. Comparison of various void fraction correlations.[10]

sinusoidal

$$q''(z) = q''_m \sin \frac{\pi z}{Z} \tag{3.21}$$

in the case considered in Fig. 3.4. q''_m in eq. (3.21) was chosen to be equal to 1.9 W/mm^2 while the system pressure is 12.3 MPa (1 Pa = 1 N/m^2) which is equal to 1785 psi.

Figure 3.4 is illustrative in the sense that it indicates the variances one may expect to encounter when different correlations are applied. In view of fuel management core analysis the reactor designer usually has experimental support for the correlation recommended. Once the void fraction is known, the coolant (and moderator) densities which are the basis for the calculation of the coolant cross-sections can be obtained.

3.6. Heat Transfer Crisis

In single-phase flow the heat transfer type is that of forced convection. In subcooled boiling or bulk boiling regimes the detachment of bubbles increases heat transfer from the rods into the coolant, but the improvement is valid only up to a certain critical heat flux where a steam layer is formed over the rod surface. In the subcooled regime the occurrence of steam film is usually local and causes a rapid temperature excursion at the rod surface, whereas a wider and smoother dryout is expected in BWRs. The vapour film insulates the rod surface, substantially reducing the heat transfer capacity. The abrupt diminution is referred to as the boiling or heat transfer crisis, and core design principles imply a clear margin to the critical heat flux which is not to be exceeded.

Due to the different flow characteristics of the PWRs and BWRs, the boiling crisis occurs in a dissimilar manner.[11] In the subcooled boiling regime of PWRs the bubble distribution is negligible, except in the immediate vicinity of the heating rod. An excessive or disturbed buildup of bubbles causes the bubble layer to grow thick and to retard, thus preventing the liquid coolant from reaching the rod surface. Boiling is changed from the so-called nucleate to film boiling and accordingly the phenomenon is known as departure from nucleate boiling (DNB).

Knowledge of the critical heat flux q''_{crit} is a cornerstone of reactor design. This quantity cannot be deduced directly, but is obtained from pertinent correlations based on experimental data. Among the best known correlations consider first the W-3 correlation[11] which is stated for a uniform heat flux $q''_{crit,e}$ along the flow direction. Nonuniform flux is accounted for by the introduction of a flux shape factor. In addition to pressure p, mass flow rate G and geometrical dimensions q''_{crit} depends on the flow quality x. A typical PWR case is studied in Fig. 3.5 where the W-3 values of $q''_{crit,e}$ are drawn[12] as a function of p and x. The negative values of the flow quality x derive from the fact that x is defined through the saturation enthalpies. From the definition of x as the ratio of vapour to mixture flow rates it follows that the enthalpy per mass unit is given by

$$H = (1-x)H_l + xH_v \tag{3.22}$$

where l and v refer to liquid and vapour, respectively. At the saturation enthalpy $H_l = H_{sat}$ and $H_v = H_{sat} + L$ where L denotes the heat of vaporization. One obtains from eq. (3.22) for the flow quality

$$x = \frac{H - H_{sat}}{L}. \tag{3.23}$$

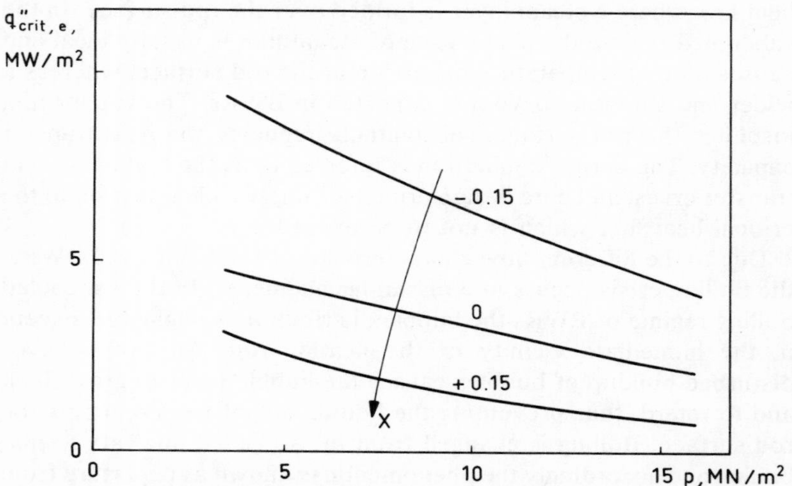

FIG. 3.5. W-3 critical heat flux as a function of pressure and steam quality.[12]

Negative values of x in Fig. 3.5 correspond to water with temperature below the saturation point.

Other correlations which are applicable within the PWR parameter range are the BW correlation[11] developed by Gellerstedt *et al.* and the Smolin[13] correlation which is used in the Soviet Union. The W-3, BW and Smolin correlations are compared in Fig. 3.6, where the results shown are for the same test problem as was used for Fig. 3.4, i.e. a sinusoidal heat flux with $q''_m = 1.9\,\text{W/mm}^2$ and $p = 12.30\,\text{MN/m}^2$. The nonuniformity of the heat flux is handled via a form factor which will be discussed later. The ordinate in Fig. 3.6 is the DNB ratio

$$\text{DNBR} = \frac{q''_{\text{crit}}}{q''}. \tag{3.24}$$

For the core considered it is found in Fig. 3.6 that the Smolin correlation yields lower DNB ratios than do the W-3 and BW correlations. In other words, the Smolin correlation is somewhat more conservative in the sense that it would predict the attainment of

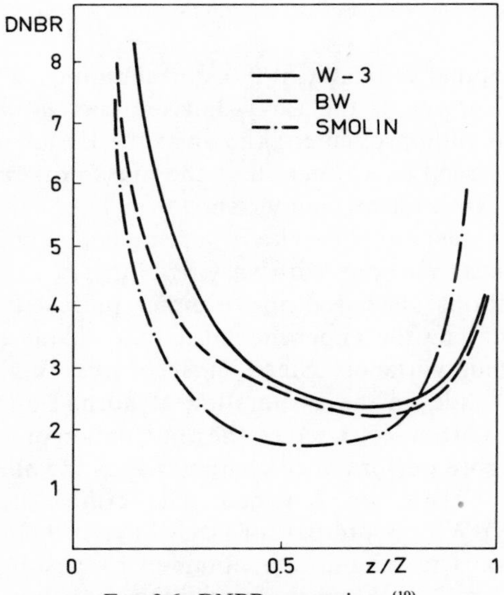

FIG. 3.6. DNBR comparison.[10]

the critical heat flux at a lower nominal heat flux than the other two would do.

In BWRs the boiling crisis is imminent in the flow regime preceding mist flow, where the vapour core of the flow channel widens to cover the entire cross-sectional area and the liquid annulus encircling the rod vanishes. The qualitative void fraction distribution preceding the dryout is seen in Fig. 3.7.

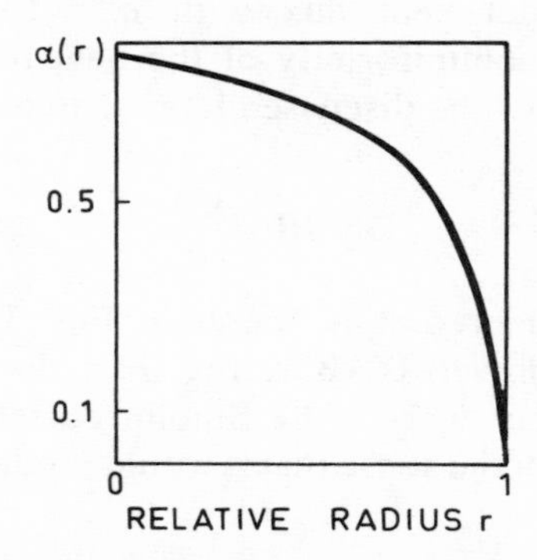

FIG. 3.7. Radial void fraction distribution in BWR flow channel.[11]

The most popular critical heat flux correlation used to predict the occurrence of dryout is the Levy–Janssen correlation,[11] together with a subsequent improvement known as the Hench–Levy correlation. It is expressed as a function of the mass flow rate G and the flow quality x in a linear manner shown in Fig. 3.8. Following an interval of low steam quality with q''_{crit} being independent of x, there appear two linear portions with varying slopes.

The correlations discussed above either represent conservative envelopes confining the experimental data available or fit the data within a certain variance. Since most of the fuel management decisions are made on the assumption of normal operating conditions, one is not often restricted by the anticipation of a boiling crisis.

In all LWR core performance characteristics the minimum values of DNBR or CHFR are specified. The critical heat flux ratio (CHFR) is the BWR counterpart of DNBR and is defined identically with eq. (3.24). The minima are obtained by a simple procedure illustrated in Fig. 3.9. When the prevailing heat flux $q''(\mathbf{r})$ and the

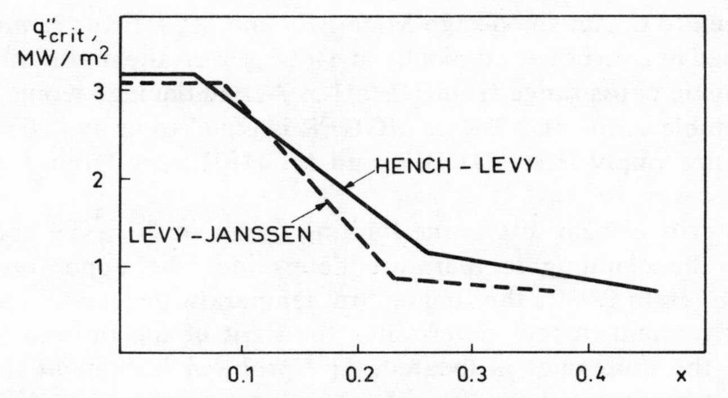

FIG. 3.8. CHF correlations.[11] Curves apply for mass flow velocity of 1350 kg/m^2/s. Reproduced from *Nucleonics* with permission of McGraw-Hill, Inc.

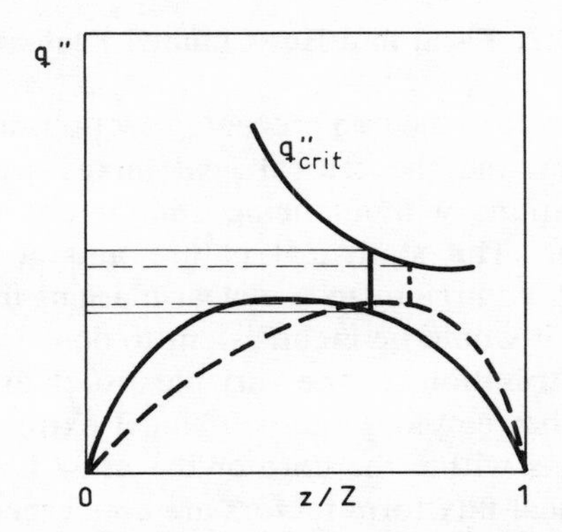

FIG. 3.9. Determination of MDNBR and MCHFR.

critical heat flux $q''_{crit}(\mathbf{r})$ are known throughout the core, the hottest channel determines the minimum DNB or CHF ratios, MDNBR or MCHFR, respectively. The criterion can be formulated as in the form

$$\text{MDNBR, MCHFR} = \min \frac{q''_{crit}(\mathbf{r})}{q''(\mathbf{r})}. \tag{3.25}$$

Figure 3.9 includes two different axial power profiles, leading one to infer that an upward shift in the maximum power reduces the MDNBR, thus reducing the margin. Due to certain transients foreseen to occur, the design MDNBRs and MCHFRs are normally specified at overpower, typically at 115% power, the minimum ratios in realistic cores range from 1.3 to 1.5. Actual damage would not be predictable until MDNBR or MCHFR is equal to unity. The design principles imply that MDNBR and MCHFR vary from 1.9 on at rated power.

The critical heat flux is the limiting factor in the axial core area where the cladding temperature determines the upper operating limit. Even in LWRs the fuel central temperature can become more restrictive and in fact determines the limit at the bottom portion where the core inlet is located. The problem is then in the heat conduction area, where linear heat rating must be limited.

3.7. Form and Hot Channel Factors

The remarks made in the two preceding sections indicate that both the void fractions and the critical heat fluxes are obtained from empirical correlations without being completely substantiated by theoretical rigour. The statistical nature and complexity of the phenomena involved urge us to use clear margins in the experimental limits. Hence it would be inconsistent to deal explicitly with the detailed local distributions of the variables, such as the heat fluxes, etc., and one rather employs ratios of local extreme conditions to average conditions within the core or the most heated channel.

As far as the heat flux form factors are concerned, the radial and axial distributions are separated by defining the radial form factor F_R as the power in hot channel divided by the power in average channel. F_R isolates the hot channel where the axial heat flux form factor F_Z is defined as the ratio of the maximum heat flux to the average heat flux. F_Z is also known as the axial hot channel factor.

Both F_R and F_Z are useful indicators of the state of the core. Minimizing F_R is tantamount to reducing the power peaking in the horizontal plane, so that none of the flow channels would be pushed far from the average. The concept of radial form factor is relevant either for a single fuel element, subassembly or rod channel. The radial flattening of power is mainly accomplished by a proper arrangement of fuel distribution.

In many cases, e.g. in PWRs, the axial power control is mainly homogeneous, leading in the crudest approximation to the chopped sine flux given in eq. (3.13). F_Z is obtained from

$$\begin{aligned} F_Z &= Z\left[\int_o^Z \sin\left(\pi\frac{z+d}{Z+2d}\right)dz\right]^{-1} \\ &= \frac{\pi}{2}\frac{Z}{Z+2d}\left[\cos\frac{\pi d}{Z+2d}\right]^{-1}. \end{aligned} \tag{3.26}$$

Upon ignoring the extrapolation distances, the sine distribution would give trivially $F_Z = \pi/2$. Acceptable values of the form factors vary greatly from one reactor type to another. The factors have to tolerate certain forms of power manoeuvring, i.e. deliberate deviations from nominal operating conditions accomplished via power control measures. 1.4 is a commonly appearing value of F_R, while F_Z is about 1.7 in LWRs. If F_R is defined assemblywise, the intra-element fine structure is accounted for by another radial factor.

References

1. El-Wakil, M. M., *Nuclear Heat Transport*, International Textbook Company, Scranton, Penn., 1971.
2. Tong, L. S. and Weisman, J., *Thermal Analysis of Pressurized Water Reactors*, American Nuclear Society, Hinsdale, Ill., 1970.
3. Hann, C. R. *et al.*, BNWL-1778, Battelle Pacific Northwest Laboratories, Richland, Wash., 1973.
4. Hewitt, G. F. and Hall-Taylor, N. S., *Annular Two-Phase Flow*, Pergamon Press, Oxford, 1970.
5. Wallis, G. B., *One-Dimensional Two-Phase Flow*, McGraw-Hill, New York, 1969.
6. Bowring, R. W., HPR 10, OECD Halden Reactor Project, Halden, Norway, 1962.
7. Rouhani, S. Z. and Axelsson, *Int. J. of Heat and Mass Transfer*, **13**, 383 (1970).
8. Levy, S., GEAP-5157, General Electric Co., San Jose, Calif., 1966.
9. Sha, W. T., *Nucl. Sci. Engng.* **44**, 291 (1971).
10. Rastas, A., VTT-YDI-4, Technical Research Centre of Finland, Helsinki, 1974.
11. Tong, L. S., *Boiling Crisis and Critical Heat Flux*, U.S. Atomic Energy Commission, Washington, D.C., 1972.
12. Eerikäinen, L., Diploma thesis, Helsinki University of Technology, 1973.
13. Smolin, V. N. and Polyakov, V. K., *Teploenergetika*, **14**, 54 (1967).

CHAPTER 4

Reactivity

INDIVIDUAL factors influencing the reactor core performance persist usually over a limited range of space and energy. Besides the violation of local design limits, each contribution must be considered globally over the entire core configuration in order to quantify the relative importance of the physical or engineering aspect concerned. The needed integral quantity or index for the global core analysis is furnished by the effective multiplication factor k which was defined in Chapter 2.

Letting the reactor system be described either by the transport or the diffusion equation, eq. (2.6) and all the conceivable approximate variants, e.g. the multigroup diffusion equation (2.52), can be written in a general form

$$B\phi = \frac{1}{k} F\phi \tag{4.1}$$

where B comprises the streaming, removal and scattering while the operator F represents the source from fission. Particular functional forms of B and F were discussed in Chapter 2.

The utility of eq. (4.1) resides in the fact that one is able to isolate the largest single eigenvalue k which is just the effective multiplication factor. From the corresponding eigenvector ϕ one obtains the neutron distribution which immediately can be inserted in

$$P(\mathbf{r}) = \sum_m e_m \int_0^\infty \Sigma_f^m(\mathbf{r}, E)\phi(\mathbf{r}, E)\, dE \tag{4.2}$$

to yield the power distribution $P(\mathbf{r})$ throughout the core. All the material properties are included in the parameters of B and F which may vary from point to point. The coolant flow condition, for example, is contained in the space-dependent multigroup absorption and scattering cross-sections.

Criticality of the reactor is expressed by the condition $k = 1$ and the disturbances are felt by the tendency to remove k from unity. The relative deviation

$$\rho = \frac{k-1}{k} \tag{4.3}$$

is known as the reactivity of the system. In operating conditions the reactor is kept critical by means of balancing the different reactivity components which separately cause positive or negative changes in the system.

4.1. Reactivity Lifetime

The most important single component of core reactivity is due to fuel burnup and depletion. Fuel is fed into and discharged from the core discontinuously with internal shuffling of the partially depleted fuel elements occurring concurrently with the reloading. Since the irradiation history of the charge is repeated for any given reload, the interval between two consecutive reloadings comprises one operating cycle.

There are two basic mechanisms by which the reactivity required for the operating cycle is inserted into the core. In the case of the natural uranium fuelled reactors CANDU and MAGNOX the high conversion ratio, i.e. the relatively high plutonium buildup, will contribute a major part of the reactivity needed to compensate the depletion. The behaviour of the effective multiplication factor k is drawn in Fig. 4.1. The abscissa in Fig. 4.1 is the fraction of the cycle length during which the reactor has been operating at full power. The cycle length is defined as the period during which the reactor can sustain criticality. The excess reactivity $k - 1$ is eliminated by reactor control devices and the reactivity lifetime T of a given loading for which $k(t) \geq 1$, $t \leq T$, is equal to the corresponding cycle length.

In view of Fig. 4.1, one should note that to achieve the initial increase in k the fuel composition must be augmented by either a graphite or a heavy water moderator system, whereas this behaviour is no more plausible in light water moderated reactors or breeders.

Besides the reactivity generated in the depletion process, there is an initial reserve stored in the core at the time of refuelling. This is manifestly important in plutonium or enriched uranium fuelled

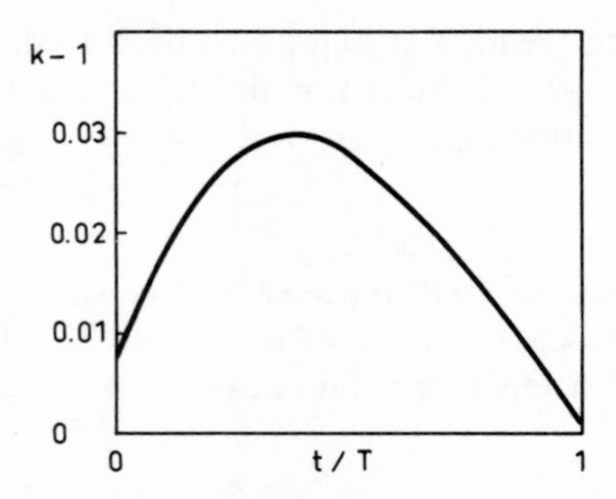

FIG. 4.1. Behaviour of eigenvalue k during the cycle of a natural uranium fuelled reactor.

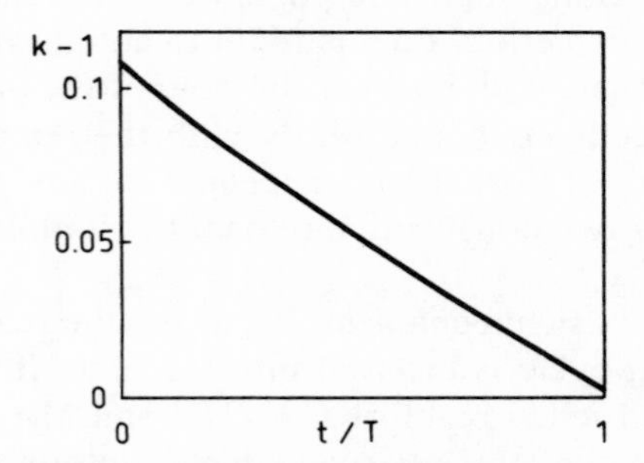

FIG. 4.2. Behaviour of eigenvalue k during the LWR cycle.

reactors where k decreases monotonically with proceeding burnup as shown in Fig. 4.2.

In thermal reactors with low initial uranium enrichment of the order of 3% the Pu contribution is still substantial, but it cannot offset the reactivity diminution caused by depletion and fission product poisoning. In breeder reactors the Pu buildup takes place in the periphery of the core and yields a negligible effect on k.

The means of reactivity control include direct insertion of absorbing materials into the core either in the form of solid control rods, burnable poisons in fuel assemblies or soluble poisons mixed with the coolant. Another important approach is to implement changes in the neutron moderation and energy spectrum.

During normal operation of a homogeneous reactor core few

strong local perturbations are present. Taking a modern PWR as an example, the main control rods are withdrawn from the core and the reactivity is adjusted by soluble poison concentration in the coolant. Variation of the moderator density is within a narrow range and the smoothness of the axial power distribution can further be maintained by proper positioning of partial length control rods. Radial distribution is flatted by loading the most reactive fuel assemblies in the outer region of the core. Even if it is somewhat idealized to claim a uniform power distribution, it can be used as a first approximation in the core analysis. In terms of independent variables associated with the core homogeneity the reactivity of an individual fuel assembly is given by the initial enrichment and the burnup exposure achieved.

Neglecting the net neutron streaming from the assembly, consider a PWR fuel element whose condition was specified above. Because of the non-leakage assumption the assembly is part of an infinite lattice and the multiplication will be denoted by k_∞. Figure 4.3 depicts k_∞ as a function of burnup for different initial enrichments. The results in Fig. 4.3 were computed by the METHUSELAH code.[1] Similar results are also given elsewhere.[2]

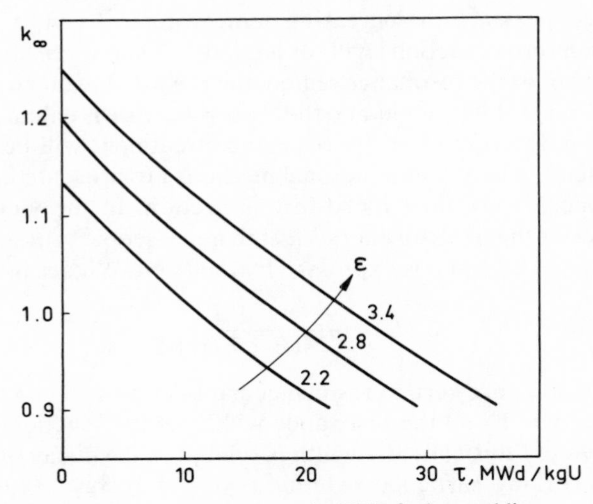

FIG. 4.3. k_∞ versus burnup for PWR fuel assemblies.

The behaviour of k_∞ can be predicted with narrow margins. This very feasibility to analyse the fuel behaviour over any given cycle is the basic incentive for reactor core fuel management.

In the simple example considered above, the desired discharge burnup of fuel determines the initial enrichment and henceforth nearly uniform neutron flux implies monotonic and smooth reactivity decrease for each fuel assembly over all residence cycles and shuffling positions. In more realistic cases and indeed in other reactors than PWRs the reactivity of a given fuel assembly depends on other facets of the core history. Changes in coolant density or control rod movements do not only influence burnup but via spectrum shifts effect Pu and fission product inventories as well.

4.2. Doppler Effect

Besides depletion, temperature dependence is most essential among the reactivity components originating directly from fuel. The foundation of this phenomenon is based on the appreciable part of absorptions neutrons undergo in resonances and on the broadening of resonance peaks with increasing temperature. Even if the total absorption cross-section will essentially remain constant when integrated over the resonance region, the remarkable flux dip already depicted in Fig. 1.14 widens and the absorption rate is enhanced.[3]

Detailed description of the resonance treatment will be deferred to Chapter 7, where computational methods are presented, and only basic concepts are introduced in this section. In the vicinity of a resonance at the neutron energy E_r the microscopic cross-section σ_i for a reaction of type i is expressed by the Breit–Wigner formula[3]

$$\sigma_i(E) = \sigma_r \frac{\Gamma_i \Gamma}{4(E - E_r)^2 + \Gamma^2} \tag{4.4}$$

where Γ and Γ_i are certain resonance parameters known as the total resonance width and the resonance width for the reaction of type i, respectively. Physically the width is related to the decay probability assigned for any particular reaction type and energy. Experimental results for the various resonance widths are included in the nuclear data compilations.[4]

The parameter σ_r is associated with the peak value of the resonance and is given by[3]

$$\sigma_r = 4\pi\lambda\!\!\!^{-2}(E_r)g_J\frac{\Gamma_n(E)}{\Gamma} \tag{4.5}$$

where Γ_n is the width of neutron emission, $\lambda\!\!\!^{-}$ is the reduced neutron wavelength and g_J is a spin factor.

Equation (4.4) is valid for both fission $i = f$ and capture $i = \gamma$, whereas scattering resonances are more complex. For neutrons with zero orbital angular momentum one has the microscopic scattering cross-section

$$\sigma_s(E) = \sigma_r\frac{\Gamma_n}{\Gamma}\frac{1}{x^2+1} + 2\sigma_r\frac{R}{\lambda\!\!\!^{-}}\frac{x}{x^2+1} + 4\pi R^2. \tag{4.6}$$

R denotes the nuclear radius and x is an abbreviation for

$$x = \frac{2}{\Gamma}(E - E_r). \tag{4.7}$$

The last term on the right-hand side in eq. (4.6) describes the nonresonant potential scattering (σ_p) upon which the resonance contribution of the first term is piled up. The second term is referred to as interference scattering and it changes sign at the resonance, causing a dip in the scattering cross-section below the resonance.

Having now introduced the pertinent notation consider the Doppler effect caused by the thermal motion of target fuel nuclei. Let the neutron velocity be denoted by $\mathbf{v}_n$ and the velocity of a fuel atom by $\mathbf{v}_A$. The relative velocity is then given by $\mathbf{v}_{rel} = \mathbf{v}_n - \mathbf{v}_A$. It is the relative neutron energy $E_{rel} = \frac{1}{2}m_n v_{rel}$ that should be used in eqs. (4.4) and (4.6). Fuel nuclei are viewed in an average sense and one introduces an effective cross-section $\sigma_i(E, T)$ where i refers to the reaction type and T denotes the fuel temperature where thermal motion is excited. $\sigma_i(E, T)$ is obtained by an equivalence procedure where one averages the cross-section $\sigma_i(E_{rel})$ over all the velocity distribution of fuel atoms. The equivalence criterion is stated through the requirement that $\sigma_i(E, T)$ must yield the same reaction rate for nuclei at rest as $\sigma_i(E_{rel})$ does for nuclei in thermal motion.

On this occasion an adequately accurate approximation is to assume that the fuel atoms obey the Maxwellian distribution

$$N^A(\mathbf{v}_A, T) = N^A\left(\frac{M}{2\pi kT}\right)^{3/2} e^{-Mv_A^2/2kT} \tag{4.8}$$

where N^A is the total fuel atom density and M the nuclear mass. Let there exist a monoenergetic neutron population with density N. The reaction rate is expressed by

$$R_i = \int v_{\text{rel}} N N^A(\mathbf{v}_A, T) \sigma_i(v_{\text{rel}})\, d\mathbf{v}_A \tag{4.9}$$

[cf. the definition in eq. (1.10)]. If the interaction is described by $\sigma_i(E, T)$ with all the N^A nuclei being fictitiously at rest, one has for the reaction rate

$$R_i = v_n N N^A \sigma_i(E, T). \tag{4.10}$$

The expression for the Doppler broadened resonance is now obtained by substituting $\sigma_i(v_{\text{rel}})$ from eq. (4.4) or (4.6) in (4.9) and setting the right-hand sides of eqs. (4.9) and (4.10) to be equal. For performing the integration over all $\mathbf{v}_A$ the most convenient variable is

$$y = \frac{2}{\Gamma}(E_{\text{rel}} - E_r) \tag{4.11}$$

whereupon the result reduces to

$$\sigma_f(E, T) = \sigma_r \frac{\Gamma_f}{\Gamma} \psi(x, \theta), \tag{4.12}$$

$$\sigma_\gamma(E, T) = \sigma_r \frac{\Gamma_\gamma}{\Gamma} \psi(x, \theta) \tag{4.13}$$

or

$$\sigma_s(E, T) = \sigma_r \frac{\Gamma_n}{\Gamma} \psi(x, \theta) + \sigma_r \frac{R}{\lambda\!\!\!^-} \chi(x, \theta) + \sigma_p \tag{4.14}$$

with the Doppler functions ψ and χ being defined by

$$\psi(x, \theta) = \frac{\theta}{2\sqrt{\pi}} \int_{-\infty}^{\infty} \frac{e^{-(x-y)^2\theta^2/4}\, dy}{1 + y^2} \tag{4.15}$$

and

$$\chi(x, \theta) = \frac{\theta}{2\sqrt{\pi}} \int_{-\infty}^{\infty} \frac{e^{-(x-y)^2\theta^2/4}\, 2y\, dy}{1 + y^2}. \tag{4.16}$$

θ denotes the relative resonance width

$$\theta = \frac{\Gamma}{\sqrt{4EkT/A}} \tag{4.17}$$

and involves the shape of the broadened resonance peak. In particular, for nuclei at rest, where $T \rightarrow 0$ and $\theta \rightarrow \infty$, $\psi(x, \infty)$ and $\chi(x, \infty)$ reduce to $1/(1+x^2)$ and $2x/(1+x^2)$, respectively. Hence eqs. (4.12)–(4.14) reduce to eqs. (4.4) and (4.6) as expected.

In order to furnish some quantitative insight into the temperature effects, consider an example involving certain lower U^{238} capture resonances for which the width parameters are given in Table 4.1.[5]

TABLE 4.1
Widths of Some U^{238} *Resonances*

E_r, eV	Γ_γ, eV	Γ_n, eV	Γ, eV
6.68	0.0250	0.0015	0.0265
21.0	0.0250	0.0090	0.0340
36.8	0.0260	0.0330	0.0590
80.8	0.0246	0.0021	0.0267
190.0	0.0220	0.1350	0.1570

A relevant measure for the strength of a resonance is the rate of reactions that neutrons undergo over the resonance peak. For this purpose the concept of resonance integral is introduced by

$$\mathrm{RI}_i^r = \int_{\Delta E_r} \sigma_i(u)\phi(u)\,du. \tag{4.18}$$

The lethargy variable [cf. eq. (1.18)] is more convenient, at least in thermal reactors, where the flux is asymptotically constant over the resonance energies. The microscopic cross-section in eq. (4.18) implies that reaction rate is counted per each individual nucleus rather than a volume element. RI is also normalized by a factor $\int \phi(u)\,du$. Even if the range of integration ΔE_r should be taken over the immediate peak it may practically be extended from $-\infty$ to ∞ because $\sigma_i(u)$ is understood to include only the peak at $u = u_r$.

There are two physical alternatives to simplify the computation of the resonance integrals in the crudest approximation. Firstly, one may consider certain resonances to be narrow in the sense that the average energy loss per collision in fuel is far less than that in moderator. The neutron spectrum in the moderator can therefore be used in evaluating resonance integrals. In contrast to this narrow resonance (NR) approximation the resonance widths can sometimes be wide enough to require the average neutron to undergo a number of collisions when passing the resonance. More realistic models fall in between the NR and the wide resonance (WR) treatments and are therefore categorized as intermediate (IR) approximation. For the resonances listed in Table 4.1 the resonance integrals are given in Table 4.2. The IR values in Table 4.2 are obtained taking an average of results from a few IR methods put forward in the literature. The geometrical situation corresponds to a LWR fuel lattice. In view of these results it is obvious that the NR approximation is accurate enough in some selected cases, e.g. the resonance at 80.8 eV, whereas it may in general yield substantial discrepancies as is emphatically seen at $E_r = 36.6$ eV.

TABLE 4.2
Certain U^{238} *Resonance Integrals*

E_r, eV	$T = 300°K$		$T = 1500°K$	
	NR	IR	NR	IR
6.68	7.246	6.355	8.157	7.024
21.0	2.859	3.091	3.306	3.564
36.8	1.814	2.615	1.981	2.917
80.8	0.361	0.361	0.531	0.530
190.0	0.168	0.275	0.193	0.319
Total	12.448	12.697	14.168	14.354

The more computational aspects of resonance absorption will be discussed in Chapter 7. It should, however, be pointed out already here that the individual resonances may interact with each other by the flux dip tails which extend to the neighbouring peak. This overlapping is considered in most calculations. Besides the resonance integral approach one may simply apply a dense multigroup discretization in energy. This is frequently done for the resonances lying in the thermal energy region, i.e. Pu^{239} resonance at 0.3 eV and Pu^{240} resonance at 1 eV.

Direct correlation of experimental results indicates a $\sqrt{T}$ dependence of resonance integrals and, in fact, the expression

$$\mathrm{RI}(T) = \mathrm{RI}(T_o)[1 + \beta(\sqrt{T} - \sqrt{T_0})] \tag{4.19}$$

is used in many occasions of overall core analysis.[3] Here the parameters $\mathrm{RI}(T_o)$ and β are specified separately for each fuel composition and geometry.

4.3. Reactivity Coefficients

Assessment of the consequences brought about by changes in core parameters is carried out by determining the corresponding effect in reactivity. Let the state of the core be specified by the reactivity ρ or by the multiplication factor k when a perturbation is introduced in the system by explicitly resetting the value of a core variable x. x may denote fuel, moderator or coolant temperatures, densities or pressure. The changes may even be an outcome of mechanical disturbances such as the bowing of a fuel element. The reactivity coefficient β_x used as a quantitative measure of the variations is defined by

$$\beta_x = \frac{d\rho}{dx}. \tag{4.20}$$

Probably the most important coefficient of reactivity is the one associated with fuel temperature. Fuel temperature variations alter the reactivity balance mainly through the Doppler effect discussed in the preceding section and, in fact, the fuel temperature coefficient is frequently called the Doppler coefficient. The most commonly used definition is

$$\beta_T = \frac{1}{k}\frac{dk}{dT} \tag{4.21}$$

which actually is in slight conflict with the definitions in eqs. (4.3) and (4.20). Strictly speaking, k^2 should appear in the denominator. In some references the definition in eq. (4.21) is multiplied by T.

Relatively large Doppler coefficients in all power reactors are an indispensable contribution to reactor stability. In low enriched thermal reactors it is mainly the U^{238} resonances which are responsible for the Doppler coefficient. With increasing burnup the buildup of Pu^{240} enhances the Doppler feedback. In plutonium-fuelled systems such as fast reactors there is an initial Pu contribution which may decay while Pu is burned. Pu^{239} resonances and especially the one at 0.3 eV behave in a complicated manner. The resonance includes a fission contribution and an increase in the Pu^{239} absorption rate could conceivably result in a positive effect of reactivity if the other contributing factors were favourable.[6] Note also that even if HTGR uses highly enriched uranium there is a sizeable amount of Th^{232} present which yields even a higher Doppler effect than does uranium.

Typical values of β_T for various reactor types are shown in Table 4.3.

TABLE 4.3
Fuel Temperature Coefficients in Certain Types of Reactor Core

Reactor type	β_T, pcm/°C
LWR	−2.5
HTGR	−3.0
LMFBR	−0.6

pcm = 10^{-5}.

The values given in Table 4.3 are only representative and will not remain unchanged if also other system parameters are varied. The Doppler coefficient of a large PWR is drawn in Fig. 4.4 as a function of fuel temperature.[7] Observe that the temperature distribution across a fuel rod is nonuniform (cf. section 3.1) and one has to estimate an effective medium temperature which yields the same Doppler effect as the realistic temperature distribution would provide.

While the Doppler coefficient is negative in all practical designs there are certain perplexities concerning reactivity coefficients derived from the coolant and moderator. Different types of reactor can hardly be discussed in the same general terms. The moderator coefficients of LWRs are examined first.

The moderator void coefficient is associated with the coolant voidage discussed in section 3.5. This is an important component in a BWR whose large negative void coefficient of reactivity β_α is utilized in axial core management. β_α varies around −100 pcm/void fraction per cent and it becomes less negative by some 50% during a typical cycle length. This is due to plutonium and fission product buildup,

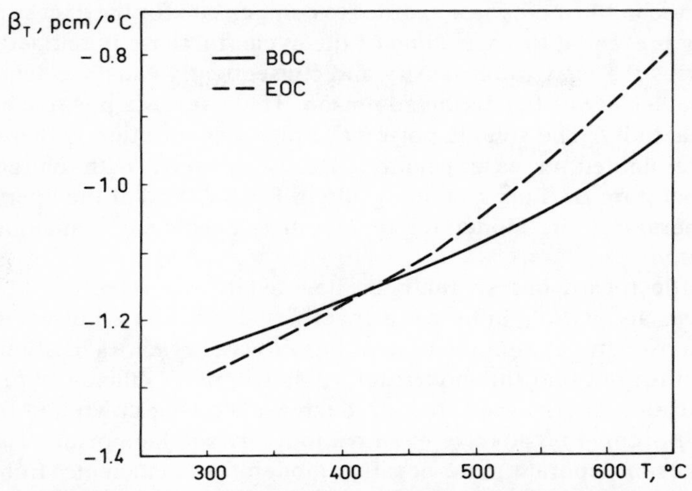

FIG. 4.4. PWR Doppler coefficient as a function of the effective fuel temperature.[7]

with increased absorptions above thermal energies. As far as the moderator temperature effect is concerned, it will be negative in BWRs over the entire temperature scale as seen in Fig. 4.5.

In certain reactors, most prominently in PWRs, soluble poison is

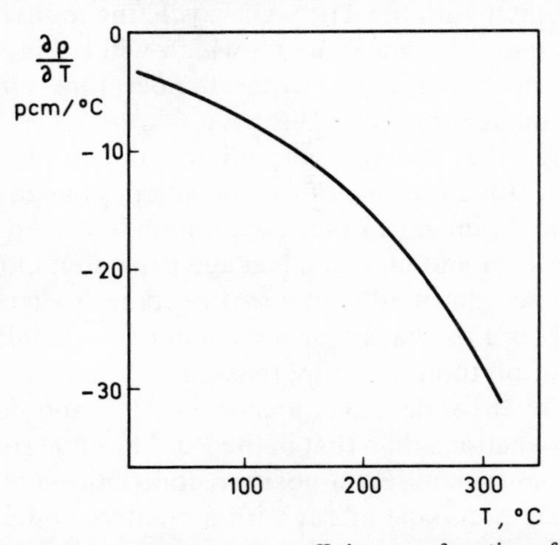

FIG. 4.5. BWR moderator temperature coefficient as a function of temperature.

injected in the moderator in order to compensate for the excess reactivity present at the beginning of the cycle. Increase in temperature reduces the moderator density and consequently causes a decrease in the density of the dissolved poison. The result is a positive effect in reactivity. The soluble poison (boron) concentration is therefore to be limited in case negative β_M is desired. With increasing temperature β_M falls monotonically in PWRs[7] and at the operating temperatures the moderator temperature coefficient is maintained negative.

Little regard has so far been devoted to the structure of fuel lattices and its role in terms of reactivity coefficients. For example, if a loose lattice, as in the case of heavy water reactors, is studied it may turn out that the moderator voids will more efficiently reduce absorption in D_2O than they do decrease slowing down and therefore β_M could be positive even without any soluble poison. Tighter lattices incorporate more negative moderator coefficients. In terms of the fuel temperature coefficient the trend is the same. Tightness implies low hydrogen to fertile material ratios. Thereby the Doppler broadening is more important.

HTGR possessing solid graphite moderator experiences little changes in moderator density and also due to the properties of helium coolant the energy spectrum is harder in HTGRs than in LWRs. Combined with the Th^{232}–U^{233} cycle the moderator produces a positive temperature coefficient[8] which will balance off a part of the Doppler coefficient. The total temperature effect, although negative, is smaller than in LWRs.

In LMFBRs the voidage of sodium coolant deserves careful consideration, since the reactivity coefficient changes sign depending on whether an inner or outer core region is voided. Voids induce a harder spectrum and enhance leakage if located close to the core exterior. The sodium void coefficient becomes negative. However, in the central core the harder spectrum entails a smaller capture-to-fission ratio in plutonium, an increase in U^{238} and Pu^{240} fast fission factors and even a decrease faster in U^{238} and fission product capture cross-sections than that in the Pu^{239} fission cross-section. All these components manifest a positive contribution and the interior voidage has an excursive effect with a positive sodium void coefficient.

There are a number of dimensional changes occurring in the core of a fast reactor which result from varying temperature expansion. Both radial and axial increments contribute to the reactivity, while the temperature gradients cause bowing of fuel elements or assemblies as a secondary effect. Inward bowing in the core centre implies a higher fuel concentration per volume element, hence increasing reactivity. These undesirable reactivity effects can be accounted for by proper design of the core.

The overall reactivity changes incurred as a result of variations in the reactor power are usually lumped together in the corresponding reactivity coefficient known as the power coefficient β_p. β_p is defined by relating the reactivity change to the percentage of the power step. Typical values of β_p range in LWRs from -2 to -7×10^{-4}/per cent in power. For example, at the cold zero power condition the core is more reactive than at the operating level. In the startup procedures the defect reactivity is displaced by reactivity control devices.

4.4. Reactivity Control

Reactivity control systems are designed with two separate objectives in mind, one corresponding to the need of compensating fuel depletion and fission product buildup and the other for the shorter-term reactivity variations due to changing power levels or other manoeuvring action. Furthermore, the control system comprises safety mechanisms adequate for initiating shutdown under all possible circumstances throughout the cycle. Different control devices also overlap with each other in order to guarantee substantial redundancies for safety and reliability.

Most versatile control of reactivity can be provided by movable control rods which both increase neutron absorption when positioned in the core and in some cases introduce substantial flux depressions at distances of a few mean free paths. The PWR fuel assembly already shown in Fig. 1.1 represents a design where the rods are spread within the fuel lattice in a subtle manner as to alleviate the immediate flux distortion. It also represents a case where the control rods are almost entirely withdrawn at nominal power and the gaps are filled with water. An opposite approach is taken in fast reactors where the control rods are just as massive as the individual fuel assemblies. In a fast core, however, the control rod effect is less localized, but extends up to the neighbouring or even more distant control assemblies, thus creating a shadow effect. Similar massive control rods are used even in certain earlier thermal reactors, e.g. a Soviet PWR design[11] where the control rod withdrawal simultaneously replaces a regular follower fuel element in the control position.

Between the two extremes of a finger-type rod cluster and a massive control assembly one encounters in BWRs an intermediate variant. Control rods are blades which can be inserted in between two fuel assemblies. Two blades are combined together in a cruciform control rod. Absorber material is usually a boron composite. In contrast to PWRs the longer-term depletion effects are covered by the BWR control rods and the withdrawal sequences are programmed prior to the operating cycle. The water spaces at the location of a withdrawn control rod introduce a thermal neutron flux peak and hence reflect to the lattice properties.[12]

Control rod movements are usually accounted for by a recalculation of the spatial neutron distribution. To evaluate the overall effect of inserting a given rod or a group of rods the control rod worth is to be known. However, the rod worths are not often employed in core management analysis, where a more detailed rod description must be adopted. The control rods perform an important function of power shaping, besides being a major device in reactivity control.

The mathematical description of control rod effects will be discussed later in Chapter 7. In lattice calculations, transport theory must be employed to describe the neutron distribution. In global reactor calculations, where diffusion theory is adequate, massive rod designs are dealt with as if being exterior regions. The extrapolation distances determined *a priori* by using accurate transport methods provide the boundary conditions. A one-dimensional cut across a control rod and an adjacent fuel assembly is shown in Fig. 4.6. The fast and thermal fluxes sketched in the figure correspond to the situation in a thermal reactor.

Instead of using the extrapolation distance condition, effective cross-sections can be derived for control rods.[12] They are determined by assuming homogeneous absorption throughout the control

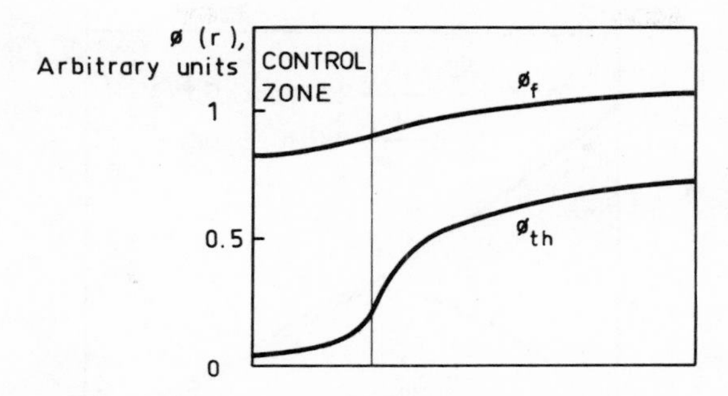

FIG. 4.6. Flux depressions due to a control rod insertion.

zone. The two methods can also be combined by letting the fast group be governed by effective cross-sections and applying the surface condition in the thermal group.

Major concern in reactivity control design must be devoted to an attempt at avoiding strong local flux and power distortions when a given control mode is coupled on. In non-saturating liquid systems soluble adjustable poisons provide an almost ideal tool in this respect. As was pointed out in the preceding section, PWRs use boron acid dissolved in the coolant. The boron concentration is reduced gradually from the initial value of about 1000 ppm to zero at the end of the cycle. The upper limit is mainly dependent on the moderator temperature coefficient which becomes less negative or even positive with increased boron concentration. Boron is used also in heavy water moderated reactors CANDU and SGHWR to control the excess reactivity.

A further means of absorber control is to load burnable poison rods in the fuel assemblies. This concept is mostly employed in LWRs where the excess reactivity is appreciable at the start of the cycle and indeed for the first core at the commencement of reactor operation. Besides having separate absorber rods, burnable poison may be dispersed in the ordinary fuel rods. Gadolinium is used in BWRs and boron rods in PWRs, the delicate difference being in the absorption cross-sections of Gd and B. The Gd^{155} and Gd^{157} have larger cross-sections which decrease faster than $1/v$, whereas $B^{(10)}$ is a $1/v$-absorber. In other words, the initial poisoning effect vanishes more rapidly in a Gd-lattice and the spectral effect does not cover

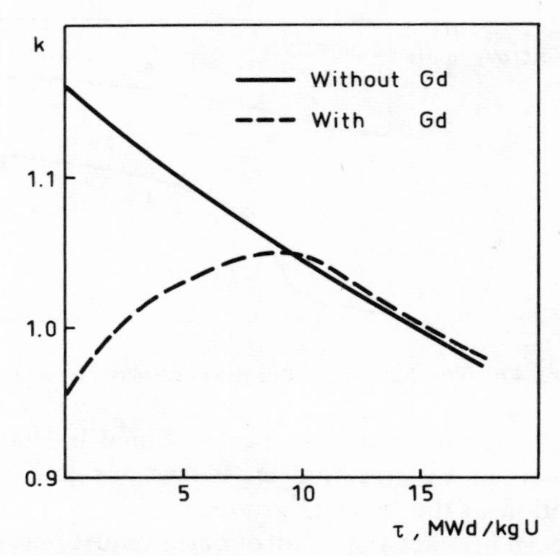

FIG. 4.7. k versus burnup for a BWR assembly with burnable poison rods.[13]

the higher thermal energies. To illustrate the burnable poison effect, Fig. 4.7 depicts the effective multiplication factor as a function of burnup in a typical BWR fuel assembly with and without Gd.[13]

The physical rationale of burnable poisons is based on the self-shielding effect. Heavy absorption at the rod surface prevents neutrons from penetrating into the rod until the surface moves in. Consider an ideal case where the burnable poison with density N^P is fully depleted over a radial depth increment dr in time dt. The decrease of poison atoms dp per unit rod length is then

$$\frac{dp}{dt} = 2\pi r N^P \frac{dr}{dt}. \tag{4.22}$$

On the other hand dp/dt must be equal to the absorption rate $2\pi rJ$ at the surface. J denotes here the surface current. Rewriting eq. (2.65) for one group surface flux $\phi = \alpha J$ one obtains from eq. (4.22) $dr = -\alpha\phi\, dt/N^P$. Letting the initial radius at $t = 0$ be denoted by r_0 and performing the integration it is found that the surface is depleted at a linear rate[12]

$$r = r_0 - \frac{\alpha\phi t}{N^P}. \tag{4.23}$$

The derivation of α encompasses the boundary condition applied at the burnable rod surface. The non-re-entrant condition of transport theory or even the asymptotic extrapolation distance will be adequate.

Instead of the control by absorbers, certain reactor types provide ample possibilities to introduce changes in neutron moderation causing a response to reactivity. Varying the coolant flow rate affects the heat transfer to the coolant and, consequently, the fuel temperature. The Doppler feedback couples on the corresponding reactivity change. In the case of the BWR the coolant flow rate is correlated to the density of the coolant-moderator. Reduced flow rate increases the steam void fraction in the core and because of the negative void coefficient the reactivity effect is negative. It is important to observe that this method is amenable to reducing reactor power only if the moderator temperature coefficient is dominant in comparison with the Doppler coefficient. Increased steam void fraction enhances the buildup of plutonium and therefore the flow control mode does not contribute excessively to the wastage of neutrons, but directs them to conversion.

The steam voids occur mostly in the upper part of the BWR core and therefore most BWR designs involve bottom entry control rods.

Other examples of reactivity control by moderation can be found in CANDU reactors where light water control zones can be inserted in the heavy water core.

Reactivity control constitutes a central part of the reactor core fuel management functions. In view of the discussion in this section it is emphatically obvious that fuel management cannot be taken too literally, but must include absorber management and coolant-moderator management as well. Moreover, fuel management must be restricted to consider not only fissile but also fertile isotopes.

4.5. Fission Product Poisoning

Fission product buildup constitutes one of the important mechanisms of reactivity distortion inherent in the fission process. At least for the purposes of thermal reactors, Xe^{135} plays the central role in the fission product inventory due both to its poisonous effect and to the proneness of inducing spatial power oscillations.

The xenon-producing portion of the pertinent fission fragment decay chain is shown in Fig. 4.8. Iodine is not actually obtained directly as a fission fragment but follows a rapidly decaying Te isotope. Most of the Xe^{135} is produced from the I^{135} branch, while the direct fission yield amounts only to a few per cent.

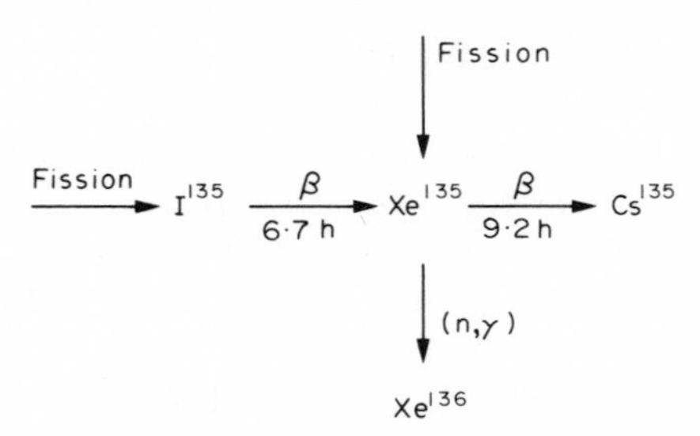

FIG. 4.8. Xe^{135} buildup chain.

After the reactor startup an equilibrium Xe concentration is reached in a few hours corresponding to the I^{135} half-life of 6.7 hr. The negative reactivity effect in LWRs is typically of the order of 3–3.5%, including Sm^{149} in addition to Xe^{135}. Also thoroughly discussed in all textbooks of reactor physics is how the equilibrium poison distribution depends strongly on the neutron flux level maintained in the core. Suppose now that the reactor is shut down. The burnout of Xe^{135} through neutron absorption will then be interrupted and the concentration begins to increase in the circumstances that prevail in power reactors. The negative reactivity effect in a BWR is shown in Fig. 4.9.[(8)]

The negative Xe shutdown reactivity reaches a maximum value in less than 10 hr and afterwards radioactive decay reduces the concentration faster than what I^{135} decay produces. Sm^{149} is nonradioactive and consequently its concentration saturates rather than peaks after shutdown.

To start up the reactor from a poisoned shutdown state it is necessary to be able to override the negative poison reactivity. This may prove to be possible only after a certain time interval corres-

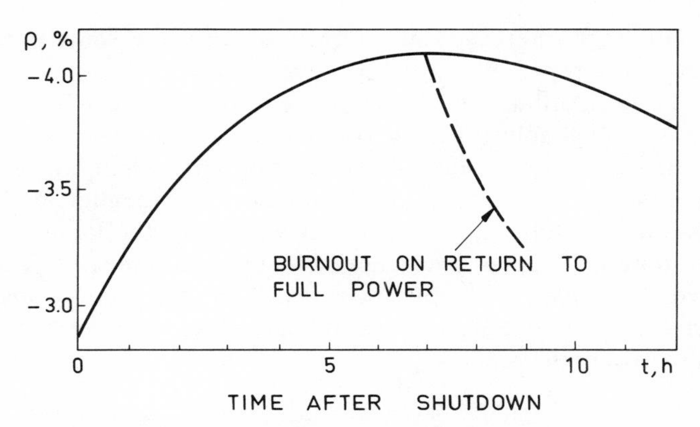

FIG. 4.9. Negative Xe^{135} reactivity buildup after shutdown.[8]

ponding to the period required for Xe^{135} concentration to be reduced sufficiently. Especially at the end of the cycle when the excess core reactivity is low, one has to bring the reactor to shutdown stepwise in order to minimize or eliminate the dead time. In case there is a reserve reactivity available and the reactor can be operated again at power, the Xe^{135} concentration can be burned out rather rapidly to the equilibrium level, as can be seen in Fig. 4.9.

The xenon transients must be considered from the viewpoint of local effects as well. The spatial variations in reactor power establish a varying equilibrium distribution of fission product poisons. Changes of the flux shape, whether due to depletion or deliberate control measures, will occasionally coincide with the existing poison distribution which may then amplify the flux variation. This interaction generates spatial oscillations.

As an illustration consider two distinct regions A and B where the thermal neutron flux is reduced in A and increased in B to correspond to a constant total power. The I^{135} distribution and consequently the Xe^{135} production will first be determined by the earlier power distribution and they would be relaxed to the new level at a rate corresponding to the decay halflife of 6.7 hr. Xenon poisoning will thus initially tend to further decrease the power in A until a certain point where the Xe^{135} concentration in B begins to work in the reverse direction. Flux is reduced in B and is correspondingly increased in *A*. Subsequently the same period of oscillations is repeated.

Xenon oscillations become imminent in large LWRs. Of the separate conceivable types of oscillations which may be encountered, the axial oscillations are important; the aximuthal ones are much less so, while radial oscillations will be negligible. Xenon oscillations are self-damped in the BWR where the large negative moderator coefficient or power coefficient provides the physical basis for strong damping of disturbances. In PWRs the xenon considerations are highly relevant and, in particular, affect the manoeuvrability of these reactors.

4.6. Reactivity Balance

In order to summarize the discussion on reactivity and to collect different factors in a proper perspective, the reactivity components of LWRs are summarized in Table 4.4. The values given are averaged from diverse reports of operation and are not representing

TABLE 4.4
Reactivity Balance in LWRs

Changes in the core	Associated reactivity %	
	BWR	PWR
1. Cold zero power to hot full power		
Fuel temperature defect	1.5	1.5
Moderator temperature and voidage	2.0	2.5
2. Equilibrium fission product poisoning	3.3	3.5
3. Burnup* compensation	6.5	10.0
4. Control margin and Xe override	1.0	1.2
5. Shutdown margin	1.0	1.0
Total	15.3	19.7

*The excess reactivity reserved for depletion depends essentially on fuel management options selected and may vary substantially.

any particular design but are rather mean parameter values of modern plants.

Comparing Table 4.4 with similar compilations for reactor types other than the LWR would most strikingly indicate the large reactivity reserve for depletion in LWRs. The relative weighting of different components is fundamental in the choice of the reactor type where the optimization of core performance simultaneously specifies the reactivity factors.

4.7. Perturbation Methods

By definition reactivity effects are tantamount to changes of the eigenvalue $1/k$ in eq. (4.1). Regarding a variety of control procedures, the reactivity insertion or withdrawal is localized within a relatively small core region while the neutron flux distribution remains almost invariable throughout the rest of the core. The problem encountered can mathematically be expressed as one to which perturbation theory lends itself.

Before proceeding to the perturbation formalism it is appropriate to introduce the adjoint transport operator $B\dagger$. Regardless of whether the transport equation or the diffusion approximation is concerned, the definition can be incorporated in an inner product form

$$(\psi, B\phi) = (\phi, B\dagger\psi) \tag{4.24}$$

where the inner product (ϕ, ψ) is equivalent to the integration over all independent variables

$$(\phi, \psi) = \int_V \int_\omega \phi(\mathbf{r}, \mathbf{v})\psi(\mathbf{r}, \mathbf{v})\, d\mathbf{r}\, d\mathbf{v}. \tag{4.25}$$

In the case of reactor physics the adjoint eigenfunction $\phi\dagger$, i.e. the solution of the adjoint equation

$$B\dagger\phi\dagger = \lambda F\dagger\phi\dagger \tag{4.26}$$

has a useful meaning of neutron importance. The argument leading to the physical contents can be found, for example, in ref. 3 where it is essentially shown that the value $\phi\dagger(r_0, v_0)$ of the adjoint function is proportional to the overall neutron flux response generated by a unit source at r_0 emitting neutrons with the velocity v_0. In other words, the relative importance of two distinct core positions r_1 and r_2 is determined by the adjoint fluxes $\phi^\dagger(r_1)$ and $\phi^\dagger(r_2)$.

Consider now the neutron balance equation

$$B\phi = \lambda F\phi \tag{4.27}$$

in either the transport or the diffusion approximation. The perturbations introduced will cause the parameters in B and F to vary such that the total disturbance influences both of them with the incremental perturbation given by ΔB and ΔF, respectively. The perturbed system is approximately governed by the equation

$$(B + \Delta B)\phi = (\lambda + \Delta\lambda)(F + \Delta F)\phi \tag{4.28}$$

where $\Delta\lambda$ represents the eigenvalue shift one is interested in. The first order approximation encompasses the assumption of an unchanged flux ϕ.

Upon neglecting the second order term $\Delta\lambda\,\Delta F\phi$ on the right-hand side of eq. (4.28) and employing the unperturbed equation (4.27) one obtains

$$\Delta\lambda F\phi = \Delta B\phi - \lambda\,\Delta F\phi. \tag{4.29}$$

Taking the inner product on both sides of eq. (4.29) the eigenvalue perturbation $\Delta\lambda = 1/k_{\text{perturbed}} - 1/k_{\text{unperturbed}}$ can be solved to yield

$$\Delta\lambda = \frac{(\phi^\dagger, \Delta B\phi) - \lambda(\phi^\dagger, \Delta F\phi)}{(\phi^\dagger, F\phi)}. \tag{4.30}$$

Assuming that a critical reactor was subjected to a perturbation the departure from criticality can be calculated from eq. (4.30) by setting $\lambda = 1$.

The applications of eq. (4.30) are manifold in the analysis of core reactivity. The computation of control rod worths is an example of the problems that can be treated by perturbation methods.

References

1. Brinkworth, M. J., AEEW-R631, Atomic Energy Establishment, Winfrith, Dorchester, 1969.
2. Henderson, R. R., in ORNL-TM-443, Oak Ridge National Laboratory, Oak Ridge, Tenn., 1974.

3. Bell, G. I. and Glasstone, S., *Nuclear Reactor Theory*, Van Nostrand Reinhold Company, New York, 1970.
4. Ozer, O. and Garber, O., ENDF/B Summary Documentation, ENDF-201, BNL 17541, Brookhaven National Laboratory, Upton, New York, 1973.
5. Höglund, R., VTT-YDI-12, Technical Research Centre of Finland, Helsinki, 1974.
6. Lunde, J. E., in *Developments in the Physics of Nuclear Power Reactors*, International Atomic Energy Agency, Vienna, 1973.
7. DOCKET-RESARA-16, Westinghouse Nuclear Steam Supply System, U.S. Atomic Energy Commission, Technical Information Center, Oak Ridge, Tenn., 1973.
8. DOCKET-STTN-50447, Standard Safety Analysis Report of General Electric, U.S. Atomic Energy Commission, Technical Information Center, Oak Ridge, Tenn., 1973.
9. Dahlberg, R. C., in CONF-720901, U.S. Atomic Energy Commission, Technical Information Center, Oak Ridge, Tenn., 1972.
10. Hummel, H. H. and Okrent, D., *Reactivity Coefficients in Large Power Reactors*, American Nuclear Society, Hinsdale, Ill., 1970.
11. Kamyshan, A. N. and Novikov, A. N., in *Reactor Burn-up Physics*, International Atomic Energy Agency, Vienna, 1973.
12. Crowther, R. L., in *Reactor Burn-up Physics*, International Atomic Energy Agency, Vienna, 1973.
13. Höjerup, C. F., RP-1-74, Danish Atomic Energy Commission, Risö, 1974.

CHAPTER 5

Reactor Operation

THE introductory discussion in the previous chapters has been confined to dealing with the aspects of core performance immediately related to fuel and to reactivity. Even if other physical facets in operating power reactors are less important for reactor core fuel management they still have relevance in providing suitable background to many of the single procedures of operation. Some of these adjacent factors will be briefly discussed in this chapter.

It is found appropriate to cover the reactivity aspects of core transients anticipated to occur relatively frequently and also the overall dynamics of plant control will be included here. Furthermore, since the core monitoring system and instrumentation partially serve the purposes of fuel management in recording the detailed operating history of the core, remarks on such a system will be made. Understanding that the on-line process control systems are tailored differently by each reactor vendor, only general observations are appropriate. For core management the utility of an on-line process computer is mainly that the information storage makes it possible to compare the realistic operating data with the predicted values which were used as the basis for the prior decisions on fuel management.

5.1. Core Dynamics and Plant Control

Many of the reactivity feedback mechanisms and control procedures as well have a prompt effect on the neutron flux and power distributions. The Doppler effect, for example, is almost instantaneous and control rods can be repositioned by gravity. The stationary diffusion formalism introduced in Chapter 2 fails there in providing a satisfactory description of the ensuing reactor transient or trip.

Figure 5.1 contains a schematic view of the various dynamic interactions usually incorporated in a time-dependent consideration of the reactor state.

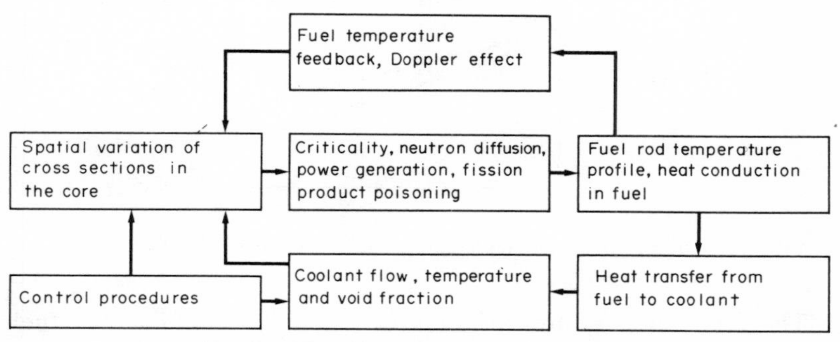

FIG. 5.1. Flow chart of reactor core dynamics.

The items appearing in Fig. 5.1 have mostly been introduced in the previous chapters. Different control procedures do not only encompass direct reactivity control but the entire plant control influences the core through this block. An illustrative example is furnished by the mechanism employed in a PWR plant which will be described below.

Suppose that a PWR power plant is designed to compensate the variations in the power demand. From the viewpoint of the reactor core the procedure commences in the parts of the plant where little concern is usually given, viz. in the pressure control of the secondary circuit. The dependence of system parameters on the reactor power is shown in Fig. 5.2.

Decreasing the pressure p_s in the secondary circuit enhances heat transfer from the primary circuit and consequently reduces the inlet core temperature T_{in}. Due to negative core temperature coefficients the reactor power P will increase to a point where the outlet temperature T_{out} and the average temperature $T \approx 1/2(T_{in} + T_{out})$ are in reactivity balance with the control reactivity. As far as the reactor is concerned, the control measures are therefore excuted automatically by the inherent core reactivity coefficients.

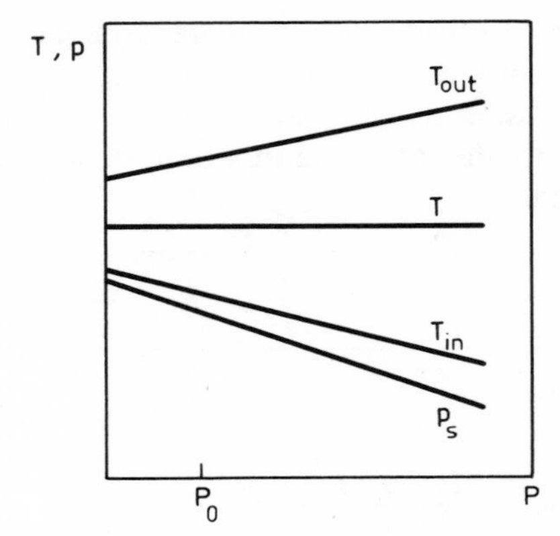

FIG. 5.2. Variation of coolant temperature and steam pressure in PWR control.

The core temperature variation in Fig. 5.2 is optimal for fuel behaviour. When the average temperature is maintained at a constant value the fuel experiences the minimal stress which results from varying temperature. Unfortunately, in practical designs the heat exchanger between the primary and secondary circuits may not tolerate the necessary temperature or pressure variation. It is clear that if only minor changes can be accommodated in p_s then increases in power must be accomplished by control rod movements and the average core temperature is increased with power.

To illustrate closer the core behaviour associated with changes in T_{in}, consider a widely used BWR benchmark problem designed for testing and comparing computational methods. With respect to the saturation temperature T_{sat}, the core inlet temperature is allowed to descend linearly during a period of 5 sec as shown in Fig. 5.3 (a).

Figure 5.3 (b) depicts the axial distribution of the surface heat flux q'' at times $t = 0$, 5.6 and 9.0 sec corresponding to the steady state, the instant of maximum fission power release, and the new temperature condition, respectively.[2] After the fuel temperature has risen appreciably enough the Doppler effect quenches the transient and afterwards a new power level will be established. In simulating the transient on the computer all the blocks of Fig. 5.1 will be coupled in and the neutron diffusion equation is used in its time-dependent

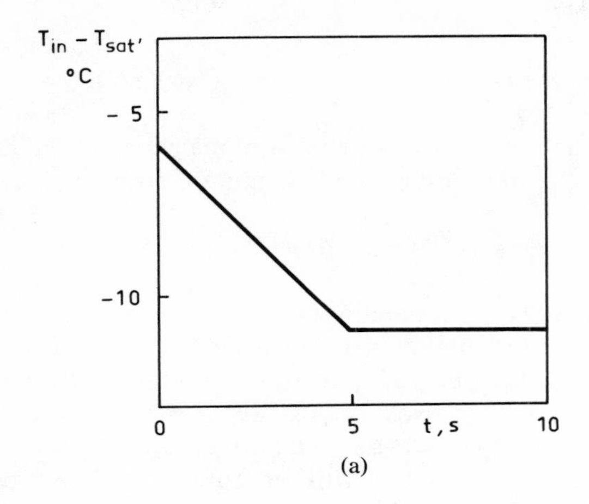

(a)

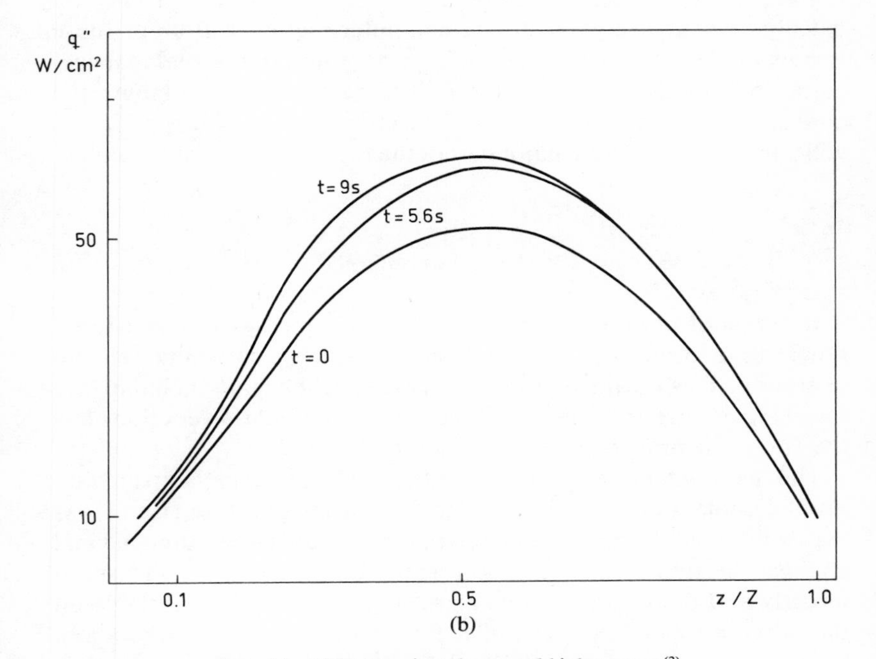

(b)

FIG. 5.3. BWR transient due to cold inlet water.[2]

form[3]

$$\frac{1}{v}\frac{\partial}{\partial t}\phi(z,t) = (B\phi)(z,t) - \lambda(F\phi)(z,t). \tag{5.1}$$

In terms of the time variable the stationary equation is solved stepwise with the derivative being computed from

$$\frac{\partial}{\partial t}\phi = [\phi(z,t+\Delta t) - \phi(z,t)]/\Delta t. \tag{5.2}$$

The time interval Δt must be chosen to be of the order 0.05–0.1 sec in order to ensure reliable results. The fission operator F in eq. (5.1) must also include the delayed neutrons[3] in addition to the prompt contribution of eq. (2.5) used exclusively in stationary analyses.

In passing it may be worth while to remark that the control problem of Fig. 5.2 is handled with an entirely different computational framework. The transients there are far less serious and can usually be analysed by real time computer methods. Both of these two examples have, however, an important link to the fuel management code modules of Chapter 7. It is customary to compute the core cross-sections separately and the cross-section block in Fig. 5.1 will later turn out to be a major part of the fuel management system.

5.2. Core Surveillance

In terms of both hardware and software core, surveillance functions serve primarily the purpose of ensuring safe and reliable operation. This means that in Fig. 5.4 the highest path including the checking of core margins and the generation of alarm functions has the foremost priority in surveillance hierarchy.[4]

The parameters to be measured include at least neutron flux profile, coolant channel flow rates, coolant temperature rise across the core, void fraction, fuel temperature and larger dimensional changes in fuel rods. Few of these variables can be measured directly and therefore on-line processing modules are inevitable on the software side. For example, flux distributions are frequently monitored by employing self-powered nuclear detectors which incorporate the capability of prompt response and facilitate the

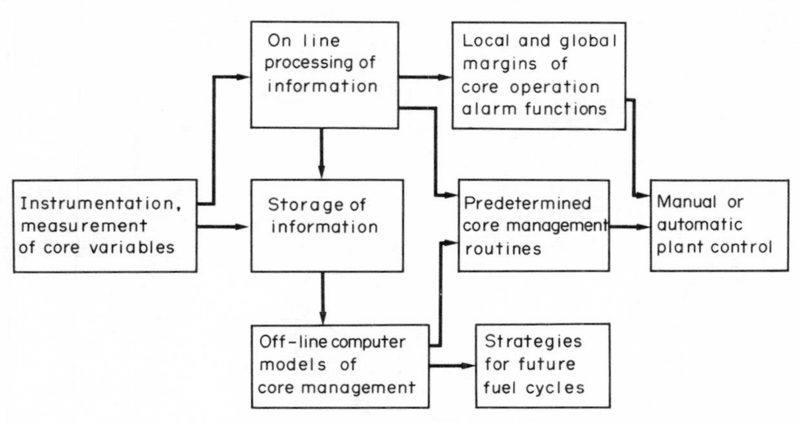

FIG. 5.4. Core surveillance and control.

measurement of nonstationary flux distributions.[5] The current generated in these detectors is due to electrons released as a consequence of neutron capture. The processing of the electron current data deals with resonances of the primary absorber and it has to account for burnup dependent properties of the detector.

No single measurement of a certain variable is usually versatile enough to provide all the relevant information nor does it have the adequate redundancy. Recalling the benchmark problem of cold inlet coolant disturbances discussed in the preceding section and shown schematically in Fig. 5.3, it is obvious that neutron flux monitoring would be opportune for initiating a timely scram. The transient is rapid and the excess heat generated will be partially accumulated in the fuel increasing the temperature and partially conducted and transferred to the coolant where the effect will be found after a certain delay shown in Fig. 5.5.[2]

In view of Fig. 5.5 it is obvious that monitoring of the coolant temperature reveals only a part of the power rise occurring.

The core surveillance procedures are naturally designed for both steady state and transient situations. At the normal condition the margin criteria would be set by the total power, linear heat rate and MCHFR. These same criteria will naturally be maintained during transients, but in a given core region the primary limiting criterion may be shifted to another. Besides the present margins advanced

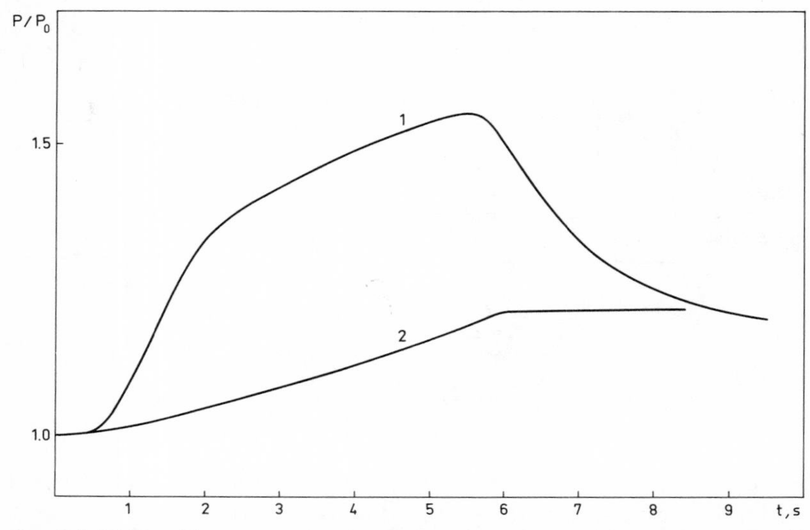

FIG. 5.5. BWR cold water transient: (1) fission power, (2) power transfer to coolant.[2]

monitoring and data processing systems are supposed to consider the trends of development in order to provide early signalling of predicted unfavourable core behaviour.

To the extent the on-line computer capacity available can tolerate it, core routines are introduced for optimizing the core performance. Eventually there may evolve an integrated on- and off-line simulation system[6] with predictive analysis capability being coupled on-line. The present book will later deal only with the off-line analysis and there the core surveillance system serves as a recorder of operating data which may be utilized in the planning of future fuel, absorber or coolant strategies.

5.3. Fuel Performance and Reactor Safety

In any treatment intersecting the nuclear reactor field one can hardly pass the question of reactor safety. The consideration of hypothetical accidents has already been included at the stage of core design where also the operating limits are specified. In LWRs, for example, the safety design incorporates requirements for the case of loss-of-coolant accidents (LOCA). Conjecturing such an accident development maximum acceptable cladding temperature is specified in the safety criteria.[7] The limitations for reactor core fuel management are based on defining an envelope of all possible operational modes that can be suggested and requiring that if LOCA commences to develop at any state within the envelope it will not result in exceeding the highest acceptable cladding temperature.

Both steady state and transient conditions are confined within the envelope. If any of the operational manoeuvres were to contain a reduction in the inlet coolant temperature, for example, then one would be obliged to have the maximum linear heat rates corresponding to Fig. 5.3 (b) as the initial values of LOCA analysis.

From the computational point of view it is worth while to point out that the core management code modules to be discussed in Chapter 7 include routines for analysis of isotopic changes during depletion. Among the output edits one can produce a fission product inventory for subsequent use in LOCA analysis. The core power distribution can be transferred from the core management modules to LOCA analysis, not to mention a few other backup services that can be provided.

In comparison with severe design basis accidents, rather frequent but many orders of magnitude smaller disturbances on fuel performance draw in fact more interest in practical core management work. To start with, consider an illustrative case study related to the LWR fuel design. The payoffs between fuel economics and reactor safety have a long history along which one phase is summarized in Table 5.1. The two different designs involved both in BWRs and in PWRs differ basically from each other only in the outer fuel pellet diameter. Within the same fuel assembly frames BWRs employ an 8×8 rod matrix as opposed to the previous 7×7. On the PWR side the corresponding shift occurs, for example, from 15×15 to 17×17. The average and maximum heat rates are reduced drastically, implying wider thermal margins to MCHFR or MDNBR, to fuel melting and clad limits in normal or LOCA conditions. On the other hand, the total number of fuel pellets or rods to be fabricated increases substantially, contributing to the increased fabrication costs.

During irradiation fuel experiences changes in density which in addition to the thermal expansion can result in unpredictable interactions between the pellets and the clad. Simultaneously the clad will lose some of its capacity to withstand strain and severe fuel failures and geometrical changes may develop.

TABLE 5.1
Design Characteristics of LWR Fuels[8, 9]

	BWR		PWR	
	7×7	8×8	15×15	17×17
Thermal power, MW	3323.0	3833.0	3411.0	3423.0
Pellet o.d., mm	12.1	10.6	9.3	8.2
Average linear heat rate, W/mm	23.3	19.8	23.1	17.1
Maximum linear heat rate, W/mm	60.7	44.0	55.4	41.0
Total number of fuel rods per core	37,400.0	49,400.0	39,400.0	51,000.0

No lucid algorithms are accessible for predicting fuel performance during irradiation. Semiempirical correlations are employed as a foundation for the rules abided by in core management. One of the major principles requires the power density to reduce throughout the residence time. The linear heat rate in the lead burnup fuel pin is qualitatively shown by the line *A* in Fig. 5.6.

The time behaviour *A* in Fig. 5.6 is recommended because it includes the monotonic decrease of pin power favourable for the stress and strain of cladding. The discontinuities between two consecutive cycles are caused by the shuffling of the fuel assemblies to new locations in the core.

A fraction of fuel pins will inevitably undergo histories of the type *B*. As will be discussed in the following chapter, the fresh fuel in LWRs is loaded in the core periphery where the neutron flux decreases steeply and the outermost rods are burned below the average during the first cycle. The clad properties can be degraded to the extent that after the power ramp of the second cycle, when the fuel expands, there will be a mechanical interaction between

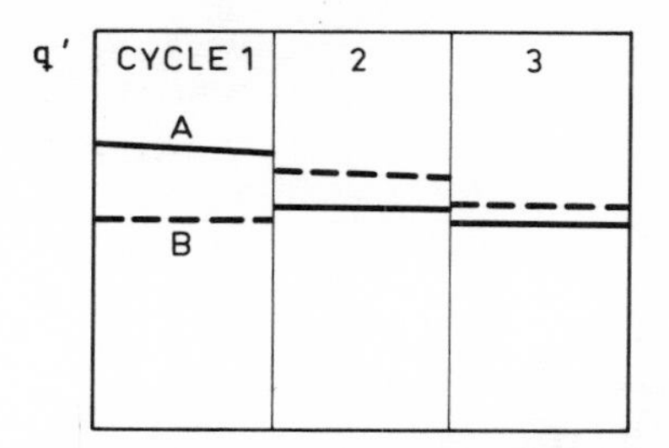

FIG. 5.6. Fuel pin heat rating during dwell time.

pellet and clad which the clad may not withstand over the ensuing irradiation period.

The problem of pellet–clad interaction is by no means limited to the beginning of the cycle. If power ramps are developed by manoeuvring during operation, the serious situation will be faced again. As a result limits based on experiments and earlier experience are set to the rates at which power can be stepped up and the magnitude of the power ramps is confined to an endurable value.

References

1. *Space Dependent Reactor Dynamics*, EUR 4731f-e, General Directorate for Dissemination of Information, Luxembourg, 1972.
2. Mannola, E., Diploma thesis, Helsinki University of Technology, 1974.
3. Bell, G. I. and Glasstone, S., *Nuclear Reactor Theory*, Van Nostrand Reinhold Company, New York, 1970.
4. *Nuclear Power Plant Control and Instrumentation*, International Atomic Energy Agency, Vienna, 1973.
5. Jaschik, W. and Seifritz, W., *Nucl. Sci. Engng.* **53**, 61 (1974).
6. Crowther, R. L. and Holland, L. K., *Proceedings of Nuclex*, Swiss Industries Fair, Basel, 1969.
7. DOCKET RM-50-1, U.S. Atomic Energy Commission, Washington, D.C., 1973.
8. DOCKET-STN-50447, Standard Safety Analysis Report of General Electric, U.S. Atomic Energy Commission, Technical Information Center, Oak Ridge, Tenn., 1973.
9. DOCKET-RESARA-16, Westinghouse Nuclear Steam Supply System, U.S. Atomic Energy Commission, Technical Information Center, Oak Ridge, Tenn., 1973.
10. Lunde, J. E., *Trans. Am. Nucl. Soc.* **20**, 237 (1975).

PART II

CORE ANALYSIS

CHAPTER 6

Variables of Core Management

IN THE approach to the principal topics of the present treatment quite a few remarks have been made concerning the restrictions imposed in making core management decisions. The discussion of this chapter is intended to point out the degrees of freedom one has to depend on in finding out feasible strategies for reactor core fuel management.

The scope of interest defined at the end of section 1.1 is still highly relevant. The standard principles and essential rules of the core management will be examined here. The succeeding chapter will deal with analytical tools devised for quantitative calculations of the core parameters associated with the basic variables. More subsequently, some methods of optimization will be discussed facilitating the final refinements in seeking the optimal fuel and longterm reactivity control strategies. It should be emphasized, however, that the search commences always with the similar screening of diverse options as the one dealt with in this chapter.

6.1. Fuel Cycle

One of the most distinctive characteristics of nuclear fuel management is the long lead time between the decision on and the practical execution of fuel loading options. The flow sheet of nuclear material shown in Fig. 6.1 indicates the principal parts of the fuel cycle. The residence time associated with each block varies from 1 up to 3 or 4 years.

If attention is paid only to the pre-core phases of the LWR fuel cycle one still has to reserve a period of about 2 years from the U_3O_8

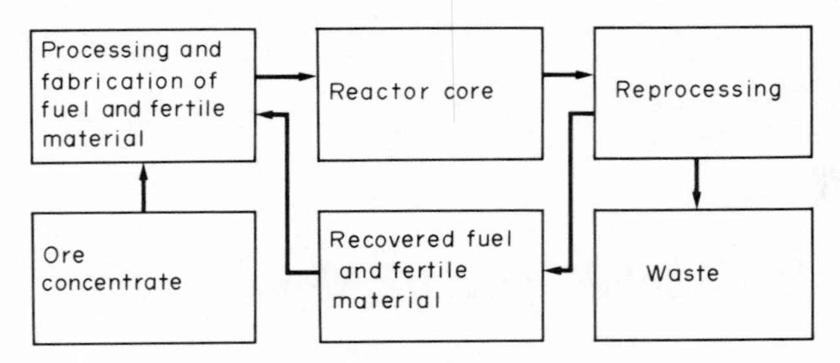

FIG. 6.1. Nuclear fuel cycle.

purchase for the diverse conversions, fuel enrichment and fabrication which take place prior to loading of the fuel into the core.

Another general remark pertains to the long core residence of fuel which implies high carrying charges relative to the cost of depletion. The flexibility available is reduced further by the fact that on-power refuelling is not feasible in power reactors with the exception of specially designed natural uranium reactors such as CANDU. The refuelling shutdown imposes essential limitations on the scheduling which can hardly be overlooked by resorting to earlier or deferred fuel loading.

Figure 6.1 provides a means of defining the scope of core management within the framework of the overall fuel cycle. Fuel management covers the technical and economic analysis and decisions by which the objective of minimizing the cost of energy production is implemented. The sub-area of reactor core fuel management would involve those tasks immediately related to the core stage. While each reactor type will have ground rules of its own, fuel management questions cannot be solved on a once and for all basis. This would be impractical because it would imply

Little consideration of future changes in energy supply and requirements, e.g. other units and future demand patterns.

No use of the accumulating data of operation usable for elimination of errors and error modes in the analysis.

No allowance for unscheduled outages due to random failure occurrences.[1]

6.2. Radial Fuelling Patterns

For both economic and physical reasons it is impractical to charge the entire reactor core at the same time and with fuel of uniform composition. Only a certain fraction of the core is replaced during each refuelling. This fraction is called batch. Moreover, the composition of the inserted fuel varies usually according to the location any specific fuel assembly or bundle is allotted to. Instead of varying the fuel composition, poison distribution could be nonuniform or these two factors could be combined.

In the radial direction the core is frequently divided into a number of concentric zones. Each zone possesses its own specification on fuel, number of batches and distribution of reactivity control. An example of the pattern of zoned multiple batch loading is shown in Fig. 6.2. The core is allowed to have hexagonal fuel assemblies in

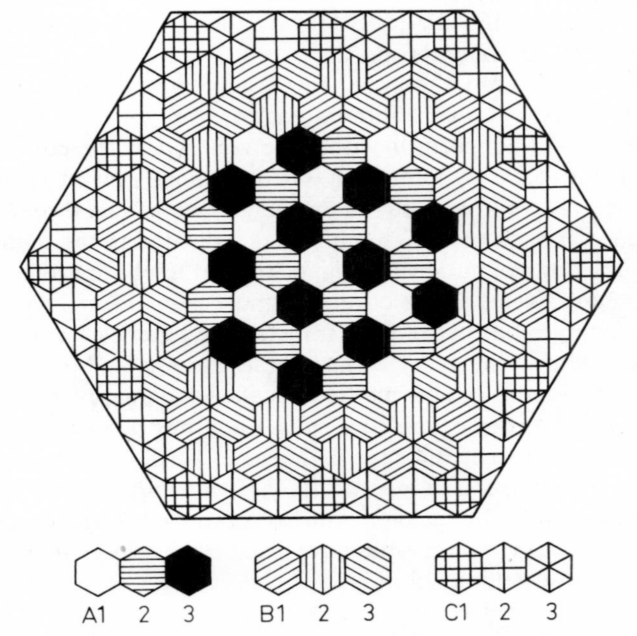

FIG. 6.2. Zonewise scattered multiple batch loading.

order to recognize that this particular refuelling scheme is or will be[2, 3] widely adopted for fast reactors. Excluding the control positions, the example core of Fig. 6.2 consists of 126 fuel assemblies grouped in three equal batches 1, 2 and 3. The three zones A, B and C include 36, 54 and 36 assemblies, respectively, corresponding to the batch sizes of 12 or 18.

At a given refuelling the 42 assemblies Ai, Bi and Ci are replaced by new ones whose composition varies from zone to zone. The index i rotates as 1, 2, 3, 1... for consecutive loadings. The feed enrichment, i.e. the ratio of fissile to total U + Pu, increases zonewise in the outward direction.

In general the annular zoning is introduced for the purpose of flattening the radial power distribution throughout the core. This is in order to maintain wider margins to fuel melting temperature or other operational limits without cutback in power. The roundelay or scattered pattern within zones reduces the power peaking locally. For example, for the SUPERPHENIX design two-zone and two-batch loading has been proposed.[2]

Once charged into the core, fuel assemblies occupy the same position throughout their entire in-reactor dwell time, i.e. there is no shuffling of fuel between or within the zones. For the core already shown in Fig. 6.2 a hypothetical zoned loading is given in Fig. 6.3. There each zone corresponds to one single batch. The three batches 1, 2, 3 consist each of 42 assemblies, as indicated in the figure. The idea of utilizing the pattern of Fig. 6.3 would imply that the batches 1, 2 and 3 are cycled in the core. The most plausible scheme would be to discharge the fuel in 3, to transfer the batches 1 and 2 to the position held previously by 2 and 3, respectively, and to charge the unirradiated fuel in 1. This method is called the out–in scheme. The fresh fuel at the edge of the core causes a strong flux gradient at the core boundary, enhancing neutron leakage. At the centre where the neutron importance is large the fresh fuel would cause a strong power peak while the leakage would be reduced. Certain reports[4] indicate that the in–out schemes have also been explored in the search for practical refuelling solutions.

Returning to the pattern of Fig. 6.2, which seemingly is of greater practical importance, one may ask if it is necessary to have the same number of scattered fuel batches in each zone. Suppose that in the

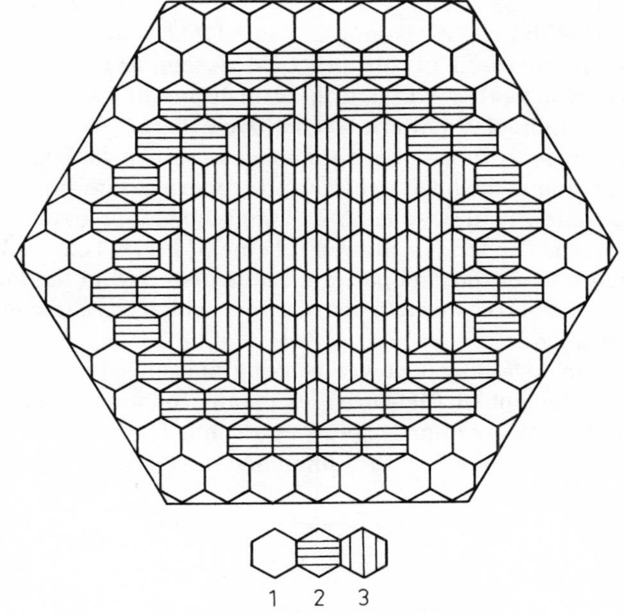

FIG. 6.3. Zoned loading.

same core two distinct batch multiplicities were desired, then one would encounter difficulties in attempting to make the refuelling times coincide. This would be no problem in reactors, where the refuelling can be carried out at full power.

In the two idealistic examples given, as well as in the subsequent patterns, the positions of control assemblies or rods should be included. This will distort the simple zone and scatter patterns in realistic applications. Secondly, the locations which leave the batch sizes to include fractions of integers must be handled separately. The central assembly in Figs. 6.2 and 6.3 would therefore constitute a batch of its own or in Fig. 6.2 the batch A1 should have 13 instead of 12 assemblies. In the case of fast reactors the fuel loading schemes are to be extended to incorporate the blanket as well. There again one can have inner and outer blanket zones with multiple batch charging.

Abandoning the zoning in Fig. 6.2 would simply mean that the zone indices A, B and C are deleted. The resulting pure roundelay is not used in any practical design. HTGR would approach nearest to that, viz. speaking in terms of radial refuelling regions the roundelay pattern is employed there.[5] A modification from the three- to four-batch roundelay is shown in Fig. 6.4. The central position is included, whereas the outermost corner elements are not counted in this scheme encompassing four batches of 30 elements each. The strict scattered scheme of Fig. 6.4 can usually be maintained only in the fuel pattern, but is distorted in the burnable poison or control rod strength distributions which have to contribute in a zoned manner.

In addition to reduced power peaking and closely related to it, partial replacement of fuel provides means for increasing the exit burnup and therefore improves fuel economy. This can qualitatively be found by considering the simplified linear excess reactivity

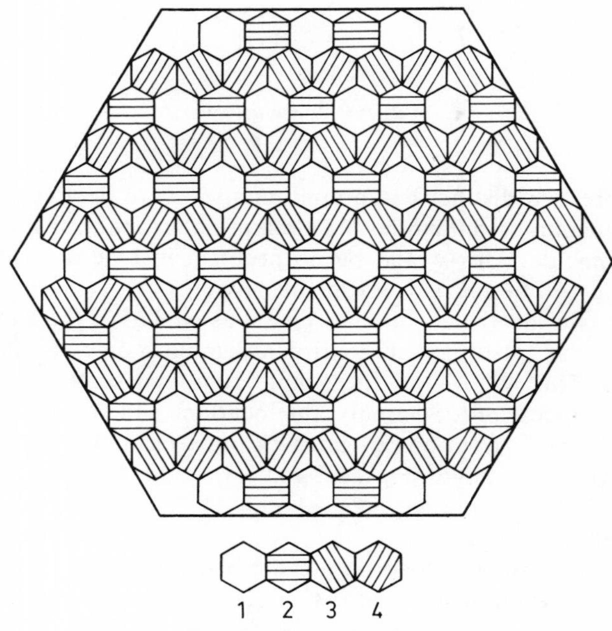

FIG. 6.4. Scatter loading.

behaviour shown in Fig. 4.2. If there were only one single batch in the core the reactivity would reduce with proceeding burnup as

$$\rho_1(\tau) = \rho_1(0)(1 - \tau/\tau_1) \tag{6.1}$$

where τ_1 denotes the discharge burnup and $\rho_1(0)$ denotes the initial excess reactivity to be reserved. In order to demonstrate the corresponding reactivity behaviour in a multiple batch operation, a transition phase from the first cycle has to be gone through.[6] τ will now denote the lead burnup and $\rho_M(\tau)$ the corresponding core excess reactivity for an M batch scatter scheme with periodic refuelling and the specified exit burnup τ_M. At the beginning of the reactor operation, ρ_M is identical with ρ_1 until the end of the first cycle

$$\rho_M(\tau) = \rho_1(\tau). \qquad \tau < \tau_M/M. \tag{6.2}$$

With one Mth part of the fuel replaced, the rest of the initial charge obeys the $\rho_1(\tau)$ dependence, whereas the fresh portion contributes to the total reactivity as $\rho_1(\tau - \tau_M/M)$. Consequently,

$$\rho_M(\tau) = \frac{M-1}{M}\rho_1(\tau) + \frac{1}{M}\rho_1(\tau - \tau_M/M), \tag{6.3}$$

where $\tau_M/M < \tau < 2\tau_M/M$.

Continuing in this manner one has

$$\rho_M(\tau) = \frac{M-m}{M}\rho_1(\tau) + \frac{1}{M}\sum_{\mathrm{i}=1}^{m}\rho_1(\tau - \mathrm{i}\tau_M/M) \tag{6.4}$$

for $m\tau_M/M < \tau < (m+1)\tau_M/M$ and $m \le M - 1$.

Finally, when $m = M - 1$, an equilibrium is reached and the reactivity obeys the relation

$$\rho_M(\tau) = \frac{1}{M}\sum_{\mathrm{i}=0}^{M-1}\rho_1(\tau - \mathrm{i}\tau_M/M), \quad \frac{M-1}{M}\tau_M \le \tau \le \tau_M. \tag{6.5}$$

Note that eq. (6.5) is valid for any subsequent cycle, i.e. $\mathrm{i} = 0$ for the lead burnup batch and $\mathrm{i} = M - 1$ for the batch least irradiated.

At the end of each cycle (EOC) $\tau = \tau_M$ and the excess reactivity goes to zero, i.e.

$$\sum_{\mathrm{i}=1}^{M}\rho_1(\mathrm{i}\tau_M/M) = 0 \tag{6.6}$$

EOC.

Upon substituting the linear expression of eq. (6.1) into eq. (6.6) one obtains

$$\frac{\tau_M}{\tau_1} = \frac{2M}{M+1}. \tag{6.7}$$

Based on the simplistic arguments given above, eq. (6.7) implies that the M batch scheme yields an improvement by the factor of $2M/(M+1)$ for the discharge burnup as compared with the single batch irradiation strategy. In particular, it is observed that in the case of continuous refuelling where $M \to \infty$ the discharge burnup τ_∞ is doubled, i.e. $\tau_\infty/\tau_1 = 2$.

One of the most obvious disadvantages associated with increasing M is related to the frequency of refuelling shutdowns necessary for off-power fuelling. Observing that the excess reactivity at the beginning of cycle (BOC) is determined by

$$\rho_M^{BOC} = \frac{2}{M+1}\rho_1(0) \tag{6.8}$$

the function $\rho_M(\tau)$ is drawn in Fig. 6.5 indicating the apparent fact that the cycle lengths vary as $2/(M+1)$ as well. The shutdown penalty makes it prohibitive to employ shorter cycle lengths than approximately one year in off-power fuelled reactors. This results in four

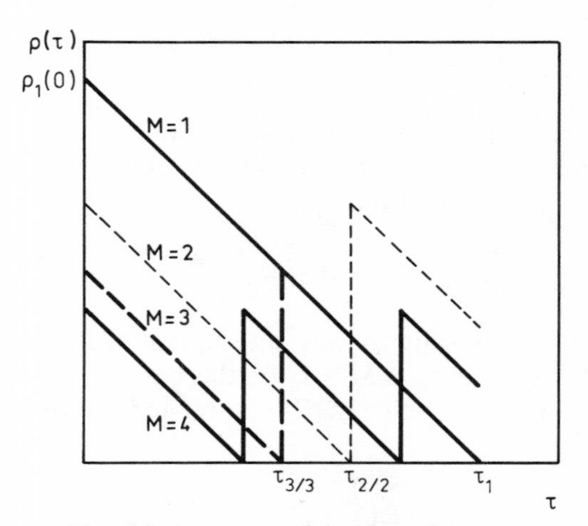

FIG. 6.5. Excess reactivity versus burnup.

batches at most, even if it cannot be conclusively inferred on the basic of the examples of Fig. 6.5.

Returning now to the original theme of radial loading patterns, there is still left at least one combination of the strict out–in and scatter schemes which were shown in Figs. 6.3 and 6.4, respectively. In order to exploit simultaneously the local and global power flattening as well as the increased discharge burnup made possible by the two schemes, a modified out–in scatter pattern is recommended for LWRs. Figure 6.6 depicts a three-batch out–in scatter

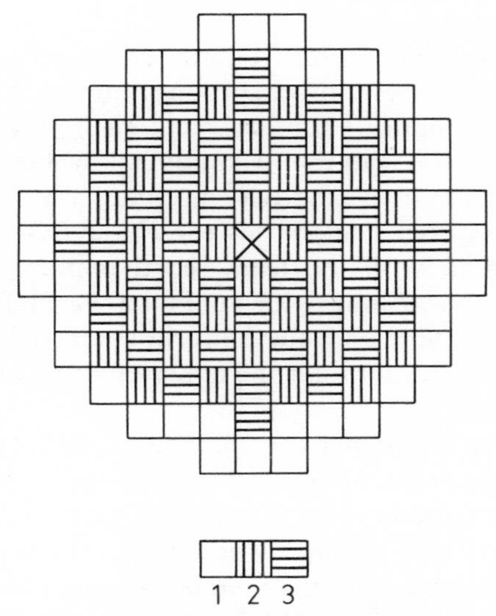

FIG. 6.6. Out–in scatter loading.

loading pattern. Excluding the central assembly, the 120 positions are grouped in equal batches. The fresh fuel is designated batch 1 and loaded at the edge of the core. At refuelling the batches are rotated as 1-2-3 with the two irradiated batches forming a checkerboard pattern over the interior core zone. Three-batch out–in scatter schemes are frequently used in PWRs while BWRs employ usually a four-batch scheme, i.e. the interior scatter region includes three batches. Note again, however, that the pattern shown in Fig. 6.6 as well as the previous schemes should be looked upon as only indicating the logic of various refuelling and shuffling methods. In practice certain deviations from the idealistic patterns are inevitable.

While the optimum radial power shape does not concern a single loading only, but involves a sequence of cycles, the optimum search is usually split into individual cycles at least in the crudest approximation. The problem boils down to the single-cycle values of the radial peaking factor F_R (cf. section 3.7) and the discharge batch burnup τ_M. Recalling that both of these parameters have upper design limits, their consideration is slightly different in PWRs and BWRs in spite of the similar loading patterns.

In a PWR where no control rod strategies are usually available due to the single mode of uniform boron control, the flattening of the power distribution appears to be attractive.[7] In fact, given the fuel assemblies of Fig. 6.6 with the maximum allowable F_R it is obvious that the longest possible reactivity lifetime is associated with the particular loading that had the lowest F_R at BOC. Conversely, a given cycle length with the least reactive practical fuel charge corresponds to the loading with minimum F_R at BOC. The BOC condition is compatible with the uniform control. The drawback of the power flattening criterion is the omission of succeeding cycles in the consideration.

In case nonuniform radial control is used, then one can resort to the degrees of freedom available for keeping F_R higher and maximizing the discharge burnup. A different view is embodied in the criterion of Barth and Haling[8] who suggest the principle of maintaining throughout the cycle the unique constant power shape that is equal to the distribution of fuel exposure at EOC. The Haling principle approach is used frequently in BWRs.

6.3. Axial Direction

The preceding section was a good demonstration of the degrees of freedom the reactor operator is allowed within the design specifications on fuelling. In the axial direction the LWR and fast reactor core designs are based on full-length fuel assemblies and therefore the loading pattern can by no means be altered. The full-length assemblies may still have an axially varying isotopic fissile composition, which leaves an option for fuel management.

For the reasons of added fabrication cost, LWR fuel rods are mostly uniform in axial fissile distribution. The axial direction is handled by poison and coolant management rather than by introducing different axial enrichment zones. As far as fast reactors are concerned, the core portion of the assemblies is usually uniform in fuel composition. There one has to recall the inclusion of a blanket and, in fact, the axial blanket is brought in by lengthening the frame and having the rods to consist of fuel and fertile zones.

Axially divided fuel assemblies make it possible to apply some kinds of systematic principles in the axial fuel management at the fuel bundle or element level. The core designs of HTGR and CANDU involve partial length fuel assemblies.

In CANDU where the fuel bundles are pushed through the core during their residence time, bidirectional fuelling has been adopted.[9] The bundles are moved in opposite directions in adjacent channels in order to smoothen and symmetrize the axial neutron distribution. In certain units of this design the core is radially divided into two annular zones where the axial movement of fuel is different. In order to have the fuel exposed to an approximately similar exit burnup, the inner zone bundles are to be moved faster or there must be some control elements present. Axial division of the fuel would also provide the option for axial shuffling of fuel in case the push-through scheme should be prohibitive.

In reactors with full-length fuel rods there is a natural incentive to flatten the axial power shape, or more precisely the axial burnup distribution of the discharge fuel. Since fuelling itself cannot contribute to this aim, both poison or absorber and coolant-moderator management become emphatically important as has already been pointed out in section 4.4.

In large reactors where the core height extends over numerous neutron diffusion lengths, the axial leakage represents a minor problem and does not buckle the neutron distribution very much. In large PWRs, for example, one could manage well even with the normal soluble boron control which is axially uniform. Partial length control rods can be used to flatten the power at nominal operating conditions, but their utility is more essential in controlling the flux shape distorted axially by power manoeuvring. The axial distribution is inherently most uneven at BOC and becomes almost uniform towards EOC. The central part of the core is depleted faster and the maximum power level glides towards the core ends where lower burnups have been experienced.

In BWRs the objective of power flattening is observed in the programming of control rod insertion and withdrawal sequences together with the coolant flow control alternative as discussed in section 4.4.

The tendency to reduce axial power peaking is taken into account not only during normal operation but during manoeuvring transients as well. The axial peaking factor F_z plays a major role in terms of restricting the conceivable states to which the reactor may be taken. Besides F_z, the concept of axial offset (AO) is frequently used to measure the integral peaking effect caused by various disturbances. Cutting the reactor into top and bottom regions V_T and V_B along the horizontal midplane the axial offset is defined by

$$\mathrm{AO} = \frac{\int_{V_T} \phi(\mathbf{r})\, d\mathbf{r} - \int_{V_B} \phi(\mathbf{r})\, d\mathbf{r}}{\int_{V_T+V_B} \phi(\mathbf{r})\, d\mathbf{r}}. \tag{6.9}$$

In view of the core monitoring system AO is very usable, because the total fluxes relative to the volume integral can be measured by ex-core instrumentation, not to mention the in-core detectors which can be used to even far more detailed flux determinations.

6.4. Approach to Equilibrium

The multiple batch loading patterns discussed in section 6.2 assume the presence of previously irradiated fuel. At the reactor startup all batches are comprised of fresh fuel whose initial composition, i.e. the fissile enrichment, differs from one batch to another in order to simulate the composition of irradiated batches.

In a typical PWR first cycle loading the initial enrichment of the batches 1, 2 and 3 of Fig. 6.6 varies from 3.5% to 1.5%. In uranium fuelled reactors enrichment refers to the U^{235} content in fuel material. BWRs employ average initial enrichments a few tenths of 1% lower than PWRs. While the cycle length is about one year in LWRs, the initial BWR cycle is prolonged by some six additional months in order to increase the burnup of the first discharge batch.

Typical LWR values of discharge burnup are drawn in Fig. 6.7. Despite the fact that the idealistic zoned, scatter and out–in scatter equilibrium patterns include only a small number of batches, there still exist a large variety of fuel assemblies with slightly different irradiation status within each batch. Therefore in a realistic case the shuffling possesses more degrees of freedom than only the few different batches. The diversity can be exploited to increase the burnup and consequently the equilibrium burnups after five to seven cycles reach the maximum shown in Fig. 6.7.

Observing that the transition to equilibrium loading may cover up to one-fourth of the plant life, careful optimization is needed in attempting to reduce the fuel costs which will in any case run higher during the first cycles than they do normally.

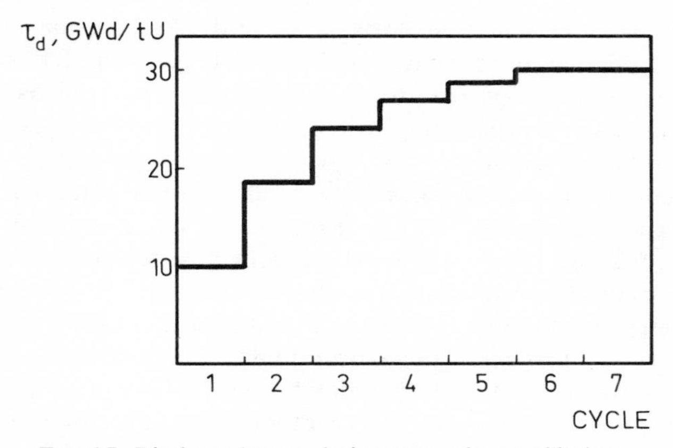

FIG. 6.7. Discharge burnup during approach to equilibrium.

Due to the burnup penalty a certain amount of compensatory excess reactivity must be reserved in the core at BOC during the approach to equilibrium. The behaviour of the excess reactivity over the first cycles is shown in Fig. 6.8. The additional reserve is based on the assumed requirement of having equal cycle lengths.

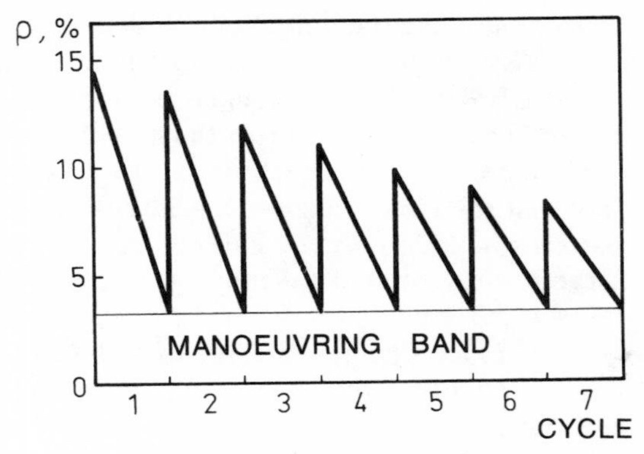

FIG. 6.8. Core excess reactivity.

The excess reactivity raises the question of counter-balancing ρ with control measures. The homogeneous soluble poison control, for example, will result in a positive moderator temperature coefficient if used excessively. Most frequently the thermal reactors involve the use of burnable poisons (cf. section 4.4) for this particular purpose of reactivity compensation as well as for the simultaneous power shaping. In fast reactors control assemblies or special diluents are in use.

To reduce the power costs associated with the initial cycles the discharged fuel assemblies can conceivably be reloaded at subsequent refuelling. The reinsertion concept has not been adopted in any substantial scale despite the potential benefits. A corresponding and even more flexible technique would be fuel sharing between a number of reactors constructed on a similar site. Partially burned fuel would be transferred to a twin unit whose refuelling is scheduled at a retarded pace with respect to the reactor where the fuel originates.

6.5. Variation of Cycle Parameters

In the idealized cases considered so far in this chapter, the cycle variables have all been fixed prior to the fuelling. Under more realistic circumstances the basis of decisions has to be evaluated in terms of variations from the assumed conditions upon which the optimization procedure had been settled.

Due to the long lead times the fuel cycle variables can conveniently be divided into two categories: the parameters related to day-by-day alternatives and the decision variables pertaining to the entire and even succeeding cycle. Movement of control rods would typically fall in the first class, whereas the second one comprises batch sizes, enrichment of the fresh batch, the location of the fresh fuel and the shuffling of the partially burnt assemblies. Moreover, the term realistic must cover unexpected failures and other random occurrences which cannot be accounted for at the *a priori* planning stage but after having taken place the associated loss has to be minimized. In other words, the decisions will be made assuming a constant or otherwise predictable capacity factor over the cycle while the factor may vary in an unexpected manner. The capacity or load factor F refers here to the ratio of the energy output (to be) generated per the maximum theoretical output at nominal power.

The significance of F is emphasized in a nuclear plant where there is a sizeable annual fixed fuel cost c_F in addition to the variable fuel cost c_V of the unit amount of energy generated. The total fuel cost c would be obtained from[10]

$$c = c_F/F + c_V. \tag{6.10}$$

The fixed component c_F includes the carrying charges of the fuel inventory whereas c_V encompasses the direct depletion. If F is now reduced by occurrences not anticipated in the fuel charge specification the fuel cycle cost increases according to eq. (6.10).

To compensate the reduction in the capacity factor the refuelling shutdown should be postponed to a date later than what was scheduled. This can be done only up to a certain deadline dictated by the power system requirement not directly correlated to the core reactivity of the specific plant with a reduced F. The shortened exposure time of the remaining fuel can be credited to subsequent refuellings where either the batch size or initial enrichment should be varied.

To study the effect of varying the batch size certain calculations have been performed in ref. 9 where the capacity factor is reduced during one single year and the batch size is modified at the following refuelling only. The planned cycle length of 6500 equivalent full power hours has been used in Fig. 6.9.[10] The discharge burnup is calculated as an average over all the fuel located in the core during the given cycle. The exit burnup decreases much faster in the case where the following refuelling is performed according to the standard recipe and there is a definite burnup gain achieved if the batch size is varied.

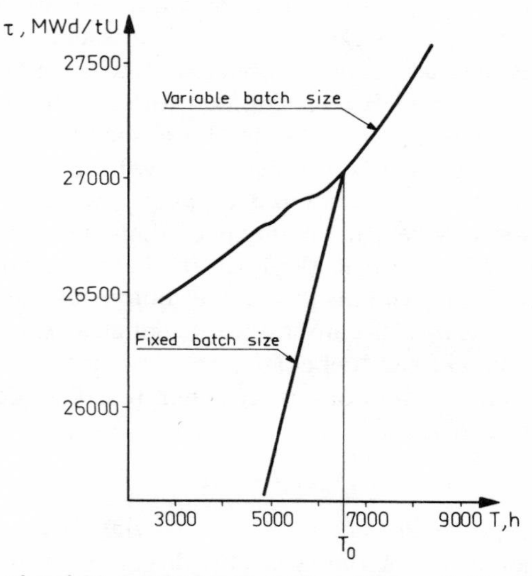

FIG. 6.9. Effect of variable batch size. Reproduced from ref. 10 with permission of the International Atomic Energy Agency.

An entirely opposite situation to the one considered above arises at EOC if the system demand should still favour operation at power instead of refuelling shutdown. Typically, this would be the case if there had occurred failures in the other power generating units and the cost of replacement energy would unexpectedly be increased about the time of the scheduled shutdown. The means to respond to the requirement of cycle extension are incorporated in the inherent core temperature defect discussed in Chapter 4 and included in Table 4.3.

Operation at reduced core temperatures beyond the scheduled shutdown is referred to as stretch-out. Figure 6.10 indicates the maximum fraction P_{max} of the nominal power P_0 that can be extracted out of the excess reactivity required ordinarily to compensate the hot full power defect.[11] In view of Fig. 6.10 it can be stated that at least 75% of the nominal power level can be produced during a stretch-out period of over 2 months. It is natural that the discharge burnup is concomitantly increased and at the same time the fixed cost per unit amount of energy becomes larger due to the reduced power level.

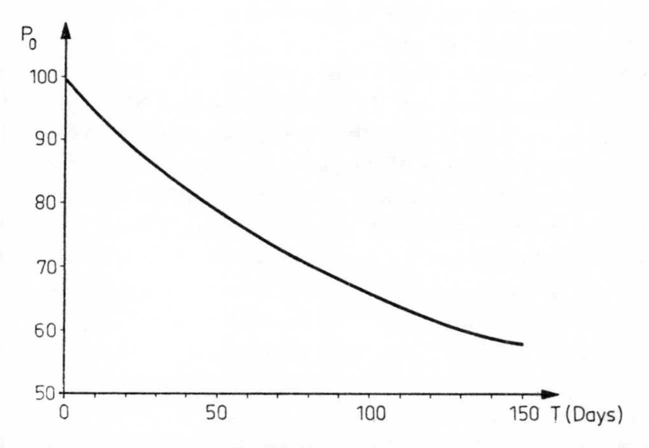

FIG. 6.10. Maximum stretch-out power. Reproduced from ref. 11 with permission of the International Atomic Energy Agency.

The long-term economic consequences of stretch-out operation must be included to cover the extra depletion of the nondischargable batches to be compensated at succeeding refuellings. Other but casual uses of stretch-out would namely require that the power peaks should be located at the fuel assemblies which will be replaced at the next refuelling. There are some further aspects of stretch-out at the end of an ordinary cycle. If the power level is reduced some weeks before the scheduled shutdown, then a larger reactivity band is available for power manoeuvring.

6.6. Coarse Breakdown of Fuel Costs

Fuel cycle costs are strongly dependent on the market situation with respect to any of the out-of-core blocks in Fig. 6.1. Until about 1990 the world fuel market will be governed by the eminence of thermal reactors and dominantly by LWRs. Even there the uranium feed will be of absolute importance until the mid-1980s when plutonium recycle is anticipated to become appreciable in thermal reactors. The above estimate is based on the assumption that fast breeders would not be installed in substantial numbers before the 1990s and the commercial reprocessing facilities would not be available until the early 1980s.

Due to the heavily increasing demand, the price of U_3O_8 will inevitably increase from the 1970 figure of about \$17.5/kg or \$8/lb by a factor of two until the end of the 1970s. Note that even the 1973 price increase of fossile fuels pushed the U_3O_8 price up by some 60–70% if a utility had unfavourable fuel purchase contracts.

Another rapidly increasing price component over the 1970s has been the cost of enrichment. The large capital investment associated with the building of new diffusion plant capacity has made it necessary to raise the price of the separative work unit (SWU) from \$26/SWU up to \$100 or even more. The same trend is followed both in the USA and USSR. The impact of centrifuge or laser enrichment technologies is still unclear.

To clarify the cost structure of uranium enrichment consider the flow rate I of the feed stream to the diffusion plant where the enriched uranium flows out at the rate P and the tails at the rate W. The total material balance requires that

$$I = P + W. \tag{6.11}$$

On the other hand, the U^{235} balance is written as

$$x_i I = x_p P + x_w W \tag{6.12}$$

where x_i, x_p and x_w denote the U^{235} concentrations of feed, product and tails, respectively. The diffusion process of isotopic separation implies a value function $V(x)$[12]

$$V(x) = (2x - 1) \ln \frac{x}{1 - x} \tag{6.13}$$

associated with a given concentration x. The separative work Δ is the difference of the total value functions

$$\Delta = WV(x_w) + PV(x_p) - IV(x_i). \tag{6.14}$$

Δ is usually expressed in separative work units (SWU) calculated per unit of product material. The ratio I/P can be solved from eqs. (6.11) and (6.12) and one obtains

$$\frac{I}{P} = \frac{x_p - x_w}{x_i - x_w}. \tag{6.15}$$

The substitution of eq. (6.11) into eq. (6.14) results in a useful relation

$$\frac{\Delta}{P} = V(x_p) - V(x_w) - \frac{I}{P}[V(x_i) - V(x_w)] \tag{6.16}$$

for separative work per unit of product. The feed enrichment x_i is the U^{235} concentration in natural uranium, i.e. 0.711; the tails assay concentration x_w can be optimized on the basis of the unit costs of SWU and feed material, whereas x_p is variable depending on the core management requirements.

The fuel cycle costs, albeit the ascendant trends of some vital components, comprise about 30% or less of the total nuclear power generating cost. A summary of the various fractional fuel costs is presented in Fig. 6.11.[12] Increasing the burnup tends to imply higher costs of fuel depletion and inventory, viz. the initial enrichment is a monotonous function of burnup. The buildup of plutonium cannot alone break this trend, since the higher initial excess reactivity must be largely controlled by parasitic neutron absorption. The fissile discharge plutonium has a reactivity value estimated at above $7/g which can be harnessed only through reprocessing. Until reprocessing services become commercially available, there will exist no unique market price *per se.*

The inventory carrying charges follow naturally the behaviour of depletion costs. The reverse trends of fabrication costs and both pre-core and post-core processing components are due to the fact that they are proportional to the flow of feed material with the ensuing $1/\tau$ dependence on burnup. The *raison d'être* of reactor core fuel management is a consequence of the existing optimal

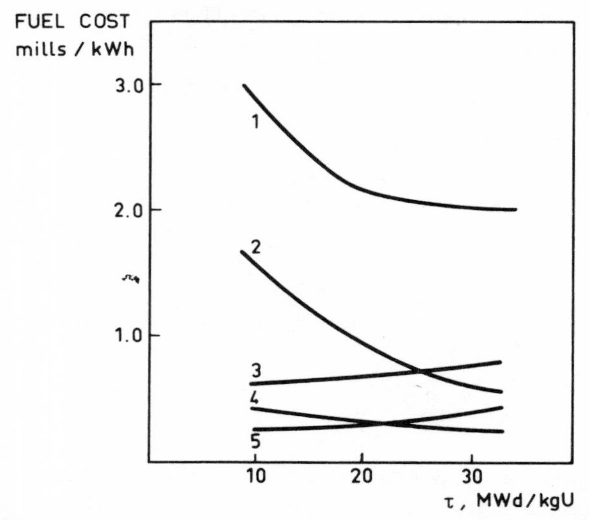

FIG. 6.11. Breakdown of fuel costs as a function of discharge burnup.[12] (1) Total fuel cost, (2) fabrication, (3) depletion, (4) processing, and (5) working capital.

solution to be explored. The complexity of the problem necessitates the use of sophisticated methods of core analysis to determine the values of meaningful core parameters in relation to fuelling strategies.

Hidden behind the economic considerations on fuel cycle there are always the fuel performance limitations that cannot be overlooked. The burnup variable in Fig. 6.11 corresponds to the average value in any discharge batch. The lead burnup assemblies have usually been irradiated to at least 10–15% higher exposure and set the limiting behaviour. Fortunately, the optimum around $\tau = 30$ GWd/tU is broad enough to alleviate severe penalties from the performance and reliability points of view.

First order calculations of LWR fuel cost can be used to evaluate the orders of different cost components.[13] A typical assessment will be made below as an example and also for the purpose of identifying the variables that can be used in cost estimates.

Consider a large PWR operated in multiple batch loading mode. The problem consists of studying the fuel cost as a function of discharge burnup τ within the interval $20 \leq \tau$, GWd/tU ≤ 40 such that equilibrium fuel cost will be calculated at 20, 30 and 40 GWd/tU. Assume also that the previous history of operation has established a relation between the average initial enrichment ϵ and the discharge burnup in the form

$$\epsilon = a_0 + a_1\tau \tag{6.17}$$

where the coefficients $a_0 = 0.8\%$ and $a_1 = 0.08\%/\text{GWd/tU}$ are chosen for this approximate computation. In order to have 1 kg of enriched uranium one needs to purchase the U_3O_8 amounts shown in Table 6.1.

TABLE 6.1
Initial Values for the Example Calculation

Discharge burnup, MWd/kgU	Initial enrichment, %	M_1 Amount of U_3O_8 purchased, kg	M_2 SWU/kgU
20	2.4	5.078	3.018
30	3.2	6.922	4.306
40	4.0	8.767	5.638

Table 6.1 is calculated using eqs. (6.15) and (6.16), observing that 1 kg of U corresponds to 1.179 kg of U_3O_8 and with the tails assay of 0.2 wt% of U^{235}.

The fuel cost will be divided into three main components: immediate fuel cost c_f, the processing cost c_p and the fuel inventory cost c_i. c_f consists of the U_3O_8 purchase and enrichment while the uranium and plutonium credits from spent fuel recovery are subtracted from c_f. The credit is calculated from the enrichment of spent fuel which is set to correspond to M_3 kg of U_3O_8 and M_4 units of separative work. The plutonium credit is determined by M_5, which is the amount of Pu contained in the spent fuel per kilogram of uranium loaded. Let the unit costs be denoted by C_i corresponding to M_i, then c_f can be obtained from

$$c_f = (C_1M_1 + C_2M_2 - C_3M_3 - C_4M_4 - C_5M_5)/\eta\tau \tag{6.18}$$

where η refers to the plant efficiency and c_f is calculated for electric power.

To carry out the example calculation it is assumed that the

discharge enrichment is independent of ϵ and τ having the concentration of 0.9%, whereupon $M_3 = 1.615$ kg and $M_4 = 0.236$ SWU. Similar assumption is made on plutonium, taking $M_5 = 6$ g of fissile Pu/kg of uranium feed. The plant efficiency is selected to be 0.33. Values of c_f for different discharge burnups are tabulated in Table 6.2. The unit costs chosen are $C_1 = C_3 =$ \$25/kg U_3O_8, $C_2 = C_4 =$ \$40/kg SWU and $C_5 =$ \$5/gPu.

TABLE 6.2
Cost Breakdown in the Example Calculation

τ, GWd/tU	c_f, mills/KWh	c_p	c_i	c
20	1.06	0.79	0.49	2.34
30	1.12	0.52	0.53	2.17
40	1.15	0.39	0.59	2.13

The processing cost c_p covers fuel fabrication C_7, reprocessing C_8 as well as the minor transportation and conversion loss expenses. Given the unit price C_6 in units of \$ per kg of uranium feed c_p is obtained from

$$c_p = C_6/\eta\tau \tag{6.19}$$

where $C_6 =$ \$125/kgU is used for Table 6.2. As pointed out previously, it is this particular $1/\tau$ behaviour of c_p that provides the major incentive to go up to higher burnups.

A precise calculation of the fuel inventory cost would have to consider detailed accounting practices and it is therefore adequate to complete the present example by making crude assumptions on the rules by which the capital is tied to the fuel. The time interval of interest is divided into three parts T_i, T_τ and T_o in Fig. 6.12. T_τ refers to the core residence time which can be calculated from

$$\tau = FST_\tau \tag{6.20}$$

where S denotes the specific power of the reactor and F is the load factor. The values $S = 30$ W(t)/gU and $F = 0.80$ are typical and they are used in the present case. T_i denotes the lead time prior to fuel loading and it is assumed in Fig. 6.12 that U_3O_8 is paid at $T = 0$ and loaded at $T = T_i$. T_o refers to the time between the discharge and selling of the spent fuel.

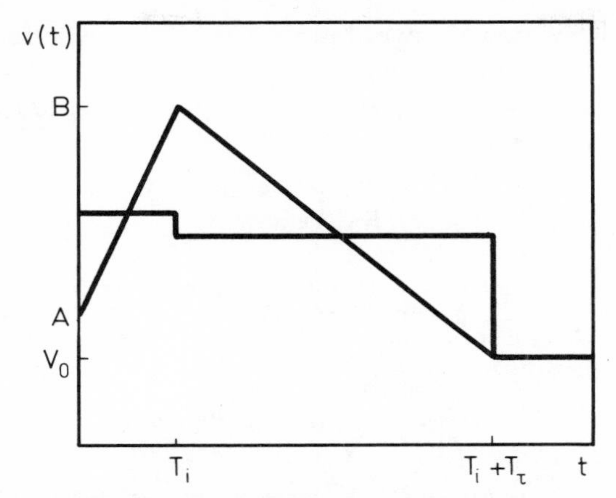

FIG. 6.12. Computation of fuel inventory cost.

Pursuant to the preceding discussion, the capital invested per kilogram of feed uranium at various times is obtained from

$$A = v(0) = C_1M_1, \tag{6.21}$$

$$B = v(T_i) = C_1M_1 + C_2M_2 + C_7 \tag{6.22}$$

and

$$V_0 = v(T_i + T_\tau) = C_3M_3 + C_4M_4 + C_5M_5. \tag{6.23}$$

The averages of $v(x)$ over the intervals T_i, T_τ and T_o are denoted by V_i, V_τ and V_o, respectively. The lengths of T_i and T_o are chosen to be one year and the fabrication cost of \$80/kgU is used for C_7.

Finally, the fuel inventory carrying charge c_i can be calculated from

$$c_i = i(V_iT_i + V_\tau T_\tau + V_oT_o)/\eta\tau \tag{6.24}$$

where the accounting is based on linear variation of the invested capital during the three intervals. i denotes the interest rate and the value $i = 0.10$ is used in computing the c_i's for Table 6.2.

Although the fuel cycle is not followed to the reprocessing stage, the pertinent charge C_8 is calculated in C_6, $C_8 = C_6 - C_7$, because it affects the price at which the spent fuel can be disposed of or sold. Note that the refuelling frequency is proportional to $1/T$ and the cost of replacement power should not be ignored in reality. The omission is in favour of lower burnups, whereas the effect would be partially offset by the reduced reliability of operation. Moreover, the highest burnup $\tau = 40$ GWd/tU could very well exceed the permissible upper burnup limit.

References

1. Critoph, E., in *Developments in the Physics of Nuclear Power Reactors*, International Atomic Energy Agency, Vienna, 1973.
2. Bussac, J. *et al.*, in the *Proceedings of International Symposium on Physics of Fast Reactors*, Power Reactor and Nuclear Fuel Development Corporation, Tokyo, 1973.
3. Cerbone, R. J., *J. Br. Nucl. Energy Soc.* **12**, 409 (1973).
4. Suzuki, Y. *et al.*, in *Fast Reactor Power Stations*, British Nuclear Energy Society, London, 1974.
5. Dahlberg, R. C. and Brooks, L. M., *Nucl. Engng. Int.* **19**, 640 (1974).
6. Lunde, J. E., in *Developments in the Physics of Nuclear Power Reactors*, International Atomic Energy Agency, Vienna, 1973.
7. Howland, H. R. *et al., Trans. Am. Nucl. Soc.* **16**, 170 (1973).
8. Barth, N. H. and Haling, R. K., *Nucleonics*, **23**, 72 (1965).
9. Pasanen, A. A., in *Experience from Operating and Fuelling Nuclear Power Plants*, International Atomic Energy Agency, Vienna, 1974.
10. Ahlström, P.-E. *et al.*, in *Peaceful Uses of Atomic Energy*, Vol. 2, United Nations, New York, and International Atomic Energy Agency, Vienna, 1972.
11. Schenk, H., in *Peaceful Uses of Atomic Energy*, Vol. 2, United Nations, New York, and International Atomic Energy Agency, Vienna, 1972.
12. Sesonske, A., *Nuclear Power Plant Design Analysis*, U.S. Energy Commission, Technical Information Center, Oak Ridge, Tenn., 1973.
13. Edlund, M. C., Lecture notes, Virginia Polytechnic Institute and State University, Blacksburg, Va., 1970.

CHAPTER 7

Computer Code Modules

TO DETERMINE the power and burnup behaviour in the reactor core constitutes a problem of remarkable dimensions. An account of the major phenomena from the level of a single fuel pin to the magnitude of the entire reactor is needed for the purpose of a reliable consideration of the relevant performance, safety and economic parameters interacting with each other. Over the past decades large investments in terms of effort and manpower have been made to establish computational algorithms to carry out the analysis. As a result almost worldwidely used standard routines have evolved for reactor core analysis. The similarity of diverse approaches is not limited to the resembling techniques of physical or numerical approximations nor to the overall organization and structure of the computer code systems, but extends to the implications caused by the rapidly developing computer hardware.

Since the objective of this entire book is based on the viewpoint of operating the reactor, it is natural to deal with a code system possessing the degree of sophistication that is likely to attract a power utility. A system implies that all the different fractions are compatible with each other. Most immediately this concerns the ratio of accuracy and computational cost. To the first order each part of the computer programme system should contribute uniformly to the total error in the results. There would be no justification in making a resonance calculation, for example, by a far more accurate and time-consuming method than what can be utilized in the subsequent lattice cell calculation or than what the measured data values would permit. Neither can the utility finally draw economic conclusions on a more refined basis than what fuel supply contracts and out-of-core cost trends dictate.

In practice the remarks made above imply that the party in charge of reactor fuel management cannot or does not have the incentive to use the methods and algorithms employed by the reactor vendor in the original design work. There exist a vast number of computer routines worked out for special and detailed purposes. Most of these are again irrelevant for reactor core fuel management. To alleviate some of the drawbacks and to increase flexibility in this respect the computer code systems are, however, designed as being modular. A module originally developed for some limited purpose can thereby be modified and combined with the other core management modules. In fact, this is done most frequently.

Due to coupled spatial and energy effects each reactor type involves certain features of its own as far as core analysis is concerned. For the sake of coherence a LWR core management system is discussed in this chapter and comments concerning other types of reactor will be presented in the next chapter. The choice is based on the LWR because this type has proliferated indisputably in larger numbers than any other type. On the other hand, the LWR core is larger in neutron mean free paths than are other cores (cf. Table 2.1). In this respect LWRs pose most difficult problems and the other types can be handled by a less heavy arsenal.

The possibility of performing a single reactor calculation that would include all the primary effects and the global core effects in one can be excluded by a simple example. In a LWR with some 50,000 fuel pins in the core a diffusion calculation should require one spatial mesh point per pin. This follows from the thermal diffusion length of neutrons being some 2 cm. Axially one would need at least 50 discretization points. Consequently, there would be a total of 2.5 million spatial points in solving the neutron diffusion or transport equation. Moreover, each point would have some five to seven independent cross-section values, fluxes and currents, not to mention the minimum of about a hundred energy groups. The rank of the ensuing linear transformation would exceed all the limits that the storage capacity of computers can sustain, whether at the present time or in the conceivable future. There would be an immense cost of running a reactor problem on the machine.

The computer code systems used for analyses of reactor core fuel management are founded on the principle of condensing the information on the energy variable while simultaneously increasing the detail with which the spatial variables are treated, commencing from zero-dimensional geometry up to the three dimensions. The temporal depletion of fuel can be included in the primary calculation or it may be executed separately.

7.1. Core Analysis System

The fundamental parts of a computational system employed in reactor core analysis are shown in Fig. 7.1. The measured nuclear data are stored in data libraries where the data are processed into a

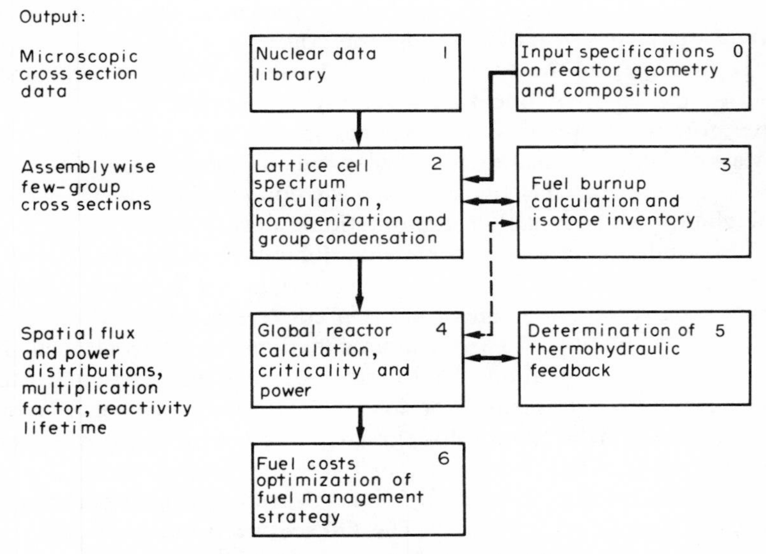

FIG. 7.1. Hierarchy of reactor calculations.

suitable format for use in lattice cell calculations. As far as thermal reactors are concerned, cell programmes may have their own multigroup (100–200 groups) cross-section libraries tested and adjusted against measurements in zero power or commercial reactors. In many cases the specific code data yield more accurate macroscopic results than do the original data available in the standard international compilations. The difference is generally small, and it has to be admitted that adjusted microscopic cross-section sets are less versatile and can be used with confidence only in the circumstances similar to those on which the modifications are based.

Block 2 in Fig. 7.1 includes the bulk of strictly physical considerations. For thermal reactors the cell spectrum codes consist of two separate parts. The fast and thermal energy ranges are analysed separately in order to account for the different neutron diffusion parameters corresponding to the slowing down and thermalization. The slowing down part is influenced by the resonance region below 10 keV, whereas neutron transport in the thermal region is governed by complex scattering interaction of neutrons with the moderator.

Lattice cell is usually made to cover fuel assembly with appropriate boundary conditions along the edges.

To homogenize the core and to condense the few group parameters at the level of one fuel assembly is relevant to the reactor operator. In reactor design and some specific other cases the fuel lattice cell can be smaller and corresponds to the unit cell. Carrying these fine structured lattice cells to basic domains in a global reactor calculation leads to higher computational costs.

The accuracy desired is somewhat difficult to define. In 1971 the International Atomic Energy Agency organized a panel[1] which drew up a set of guidelines for the predictive accuracy desired in reactor calculations. The recommendations are reproduced in Table 7.1. The tabulation is not concerned with any particular fuel management requirements, nor does it allow for any specific reactor design. Nevertheless, the values given in Table 7.1 give a fairly good description of what is desirable in LWR fuel management.

The calculations in block 2 of Fig. 7.1 are performed parametrically with proceeding burnup. The link between blocks 2 and 3 is passed recurrently such that the homogenized multigroup cross-sections for fuel assemblies are obtained as a function of burnup. Predominantly for the BWR, where the void fraction history of the assembly has influenced the neutron flux spectrum and consequently depletion, the cell burnup calculations are carried out as a function of burnup with unit weighting and void fraction weighting. Fuel temperature and moderator density are the other essential parameters used as variables in the tabulation.

Module 3 contains the algorithms required for the upkeep of the

TABLE 7.1
Estimated Power Reactor Burnup Physics Accuracy[1]

	Predictive accuracy %		Experimental accuracy %	
	Objectives	Present capabilities	Objectives	Present capabilities
Reactor thermal power	—	—	±2	±2–5
Steady state power distribution				
Within a fuel pin	±5	±10–20	±5	±5
Fuel pin relative to assembly	±2	±3–5	±2	±3
Axial, within an assembly	±3–5	±6–10	±2	±3
Radial, between assemblies	±1–3	±3–8	±2	±3
Overall, pellet to average	±3–5	±8–13	±2	±5
Steady state reactivity				
Initial neutron multiplication	±0.25–0.5	<±1		
Reactivity lifetime	±2–5	±2–10		
Fuel burnup				
Peak pellet	±3	±5	±2	±2–5
Fuel assembly	±2	±4	±2	±3–6
Discharge batch	±2	±3–5	±2	±3–6
Isotopic composition[a]				
Local (pellet)				
U^{235} depletion	±2	±5	±1	±0.5–1
Pu^{239}/U ratio	±2	±4	±1	±1–2
Net fissile atoms produced/U	±2	±4	±1	±2–3
Fuel assembly				
U^{235} depletion	±2	±5	±1.5	±1.5–2
Pu^{239}/U ratio	±2	±5	±1.5	±2–3
Net fissile atoms produced/U	±2	±5	±1.5	±3–4
Discharge batch				
U^{235} depletion	±2	±5	±1.5	±1.5–2
Pu^{239}/U ratio	±2	±5	±1.5	±2–3
Net fissile atoms produced/U	±2	±5	±1.5	±3–4

[a]Assuming that the burnup is known in predictive comparisons.

isotopic inventory of fuel composition. The depletion effects are accounted for at discrete burnup timesteps when the updating of the isotopic composition takes place as well.

All the preceding stages of the analysis are activated in building a detailed tabulation of few group homogenized cross-sections. The reactor simulation in box 4 includes a three-dimensional power distribution analysis. In PWRs where the axial distribution is fairly uniform, most of the fuel management calculations can be carried out in two dimensions with rather infrequent normalization of the axial power shape. Burnup calculation is implicit in the sense that precalculated tabulations from boxes 2 and 3 are employed. During each burnup step the core is burnt, assuming a constant flux upon which the new value of fuel burnup is based.

Fuel depletion can in principle be coupled to the reactor simulation and block 3 can be coupled to block 4. As stated previously, this would lead to a more careful and explicit consideration of the fuel burnup history. The increased computational costs can become prohibitive for this method to be used in reactor core fuel management.

The inclusion of block 5 in the system is a consequence of the thermohydraulic feedback effects playing their role in the power shaping. Besides providing the fuel temperatures and moderator void fractions, this box will automatically compute the MCHFR or MDNBR, thus providing also information on the admissibility of the reactor performance state.

To a large extent block 6 will not be discussed until Chapter 9. The box actually represents an interface between the core analysis system and the utility planning system. Therefore it is deferred to the context of the discussion on integrating the reactor unit into the power generation system.

To conclude this introductory section Table 7.2 provides an overview of the latitudes in energy and space variables at the different stages of calculations. The recipe of Table 7.2 is problem sensitive and contradictory situations may well arise.

Inspection of Table 7.2 reiterates the earlier statement on the trend. Allowing an increase in spatial degrees of freedom and an increase of dimension is counterbalanced by taking a lower number

TABLE 7.2

Description of Space and Energy Variables in LWR Core Analysis

Domain to be analysed	Box in Fig. 7.1	Energy points or groups	Dimensions	Spatial mesh points
Fuel pin	2	75–200	0–1	1–20 per pin
		15–100	1	1 per pin
Fuel assembly	2	2–15	1	1 per pin
		2–10	2	20–200 per assembly
Entire reactor	4	2	2	1–5 per assembly
		1–2	3	10–20 per assembly
Fuel strategy	6	1	0–1	20–100 per reactor

of energy groups. The condensation of information in neutron energy variable is to be accomplished in a manner where the few or one group cross-sections retain the essential contents of a multi-group structure.

7.2. Nuclear Data Libraries

Large volumes of data including up to the order of one million quantities are available in the literature. The measured data values have been or will be evaluated and compiled in computer files in a format suitable for automatic storage and retrieval.[2]

The evaluation process consists of a critical review of the measured data, the selection of the acceptable and reliable items and of averaging the different results for a given nuclear constant. Frequently, measured data points are intolerably sparse and therefore interpolation, extrapolation and theoretical models have to be applied in order to obtain consistent and satisfactory data sets.

Nuclear data compilations include practically all the neutron and non-neutron data required for fuel management. In terms of energy the range of interest lies naturally from zero energy to 12–15 MeV. Relevant neutron data comprise

Microscopic cross-sections of neutron induced reactions (cf. section 1.4).
Angular distributions of collided neutrons for both elastic and inelastic scattering.
Energy distribution of inelastically scattered neutrons.
Average number of secondary neutrons per fission.
Energy spectra of fission neutrons.
Angular and energy distributions of neutrons, protons, α-particles and γ-rays, etc., released in reactions.
Resonance width parameters.
Fission product yields and cross-sections.

The above data are given for up to 200 different elements, isotopes, molecules or specified mixtures of elements. Examples of element mixtures to be considered as independent materials are certain fission products which are lumped to form pseudomaterials.

Some of the stored nuclear data need further processing before fitting as input to box 2 of Fig. 7.1. Lattice cell programs include a multigroup data library where the cross-sections and other quantities have been averaged for. Typically the library would have some 100 energy groups and in any event noticeably less than what is the resolution in the data files. Letting $y(E)$ denote any of the microscopic data species, the averaging procedure for the energy group n is of the type

$$y_m = \frac{\int_{\Delta E_n} w(E) y(E)\, dE}{\int_{\Delta E_n} w(E)\, dE}$$

where $w(E)$ denotes a weighting function. For fission related quantities the fission spectrum is an applicable weighting, whereas the slowing down region employs usually the asymptotic E^{-1} weighting. Similarly for thermal neutrons the Maxwellian spectrum of eq. (1.20) is used.

Any function $y(x)$ is represented by tabulated values of x_i and $y(x_i)$. Furthermore, depending on the type of the variables an interpolation method is specified for covering the interval where no evaluated data points are available. In the ENDF/B library, for example, there are five schemes allowed as shown in Table 7.3.[(2)]

TABLE 7.3
ENDF/B Interpolation Schemes

Option	Function
Constant	$y(x) = A$
Linear–linear	$y(x) = A + Bx$
Linear–log	$y(x) = A + B \ln x$
Log–linear	$y(x) = A \exp(Bx)$
Log–log	$y(x) = Ax^B$

7.3. Lattice Cell Module

A flow sheet of the lattice cell algorithm is shown in Fig. 7.2. Although logically different operations occupy their own boxes in

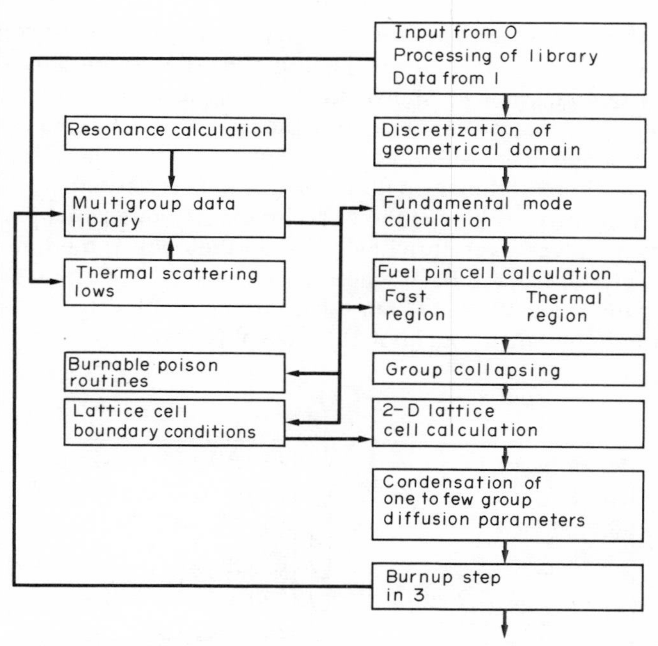

FIG. 7.2. Lattice cell program submodules (cf. block 2 in Fig. 7.1).

the figure, one should be reminded that the boxes are of largely varying importance from the overall point of view. Apart from the multigroup data library, the essential blocks in Fig. 7.2 are those corresponding to fuel pin cell or lattice cell calculations and group restructuring. The lattice cell will in this discussion be extended to coincide with a fuel assembly. Undoubtedly, one would sometimes hesitate in separating a single fuel assembly from the LWR lattice. On these occasions the boundary conditions applied along the lattice cell edge should be evaluated with special care. It can be done by a temporary extension of the lattice cell to cover regions of the adjacent assemblies.

Decoupling of individual fuel assemblies can be supported by physical arguments, viz. the thermal energy coupling extends over only the first few pins at the assembly boundary and the boundary conditions can be expected to handle the problem. PWRs have large fuel assemblies where most of the fuel cannot sense the effect of different neighbouring assemblies. BWRs employ smaller fuel assemblies, but as shown in Fig. 7.3 the fuel lattice cells are decoupled by the assembly can and the broader water spaces. There is also a noteworthy difference between uranium fuelled BWRs and PWRs in the enrichment pattern of single fuel assemblies. Due to the thermal flux peaking in the water spaces the outer fuel pins close to the gap have lower enrichment[3] than the ones in the interior. In BWRs attempts have been made to optimize the enrichment variation inside the assembly.[4] On the contrary, PWRs have a uniform initial enrichment throughout the assembly.

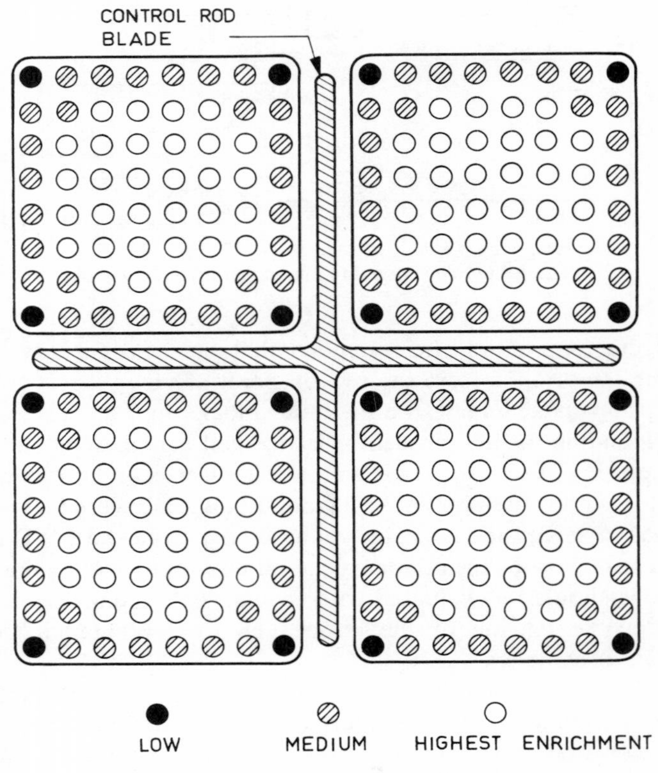

FIG. 7.3. BWR fuel assemblies.

Fast Spectrum

Fast neutrons have long mean free paths in the fuel lattice and cannot consequently feel the geometry in great detail. The spatial variable is usually treated via the fundamental mode buckling description to be introduced below. As already discussed in Fig. 1.13, there exists a substantial difference in the fast flux, depending on whether the fuel or moderator regions are being considered. The slowing down calculation must therefore allow for separate fuel and moderator regions.

The development of the computational algorithm commences from eqs. (2.18) and (2.20) where lethargy instead of energy is taken as an independent variable for the sake of convenience. Further approximations will be made in the description of neutron leakage and, on the other hand, in the slowing down scattering integrals appearing in the equation.

The description of elastic scattering is crucial in the calculation, and the slowing down approximations will pertain mostly to this contribution, whereas inelastic scattering is covered in a more straightforward manner. The basic relations of elastic scattering were given in section 1.7 and one observes in particular that the maximum energy loss per collision was found to be $(1-\alpha)E_i$ [cf. eq. (1.17)]. In terms of lethargy the final stage u_i lies within

$$u_i \leq u_f \leq u_i + \ln 1\alpha_m \tag{7.1}$$

where m refers to the target nucleus and α_m is defined in eq. (1.16). Including all the nuclear species explicitly in the scattering terms, eqs. (2.18) and (2.20) are rewritten as

$$\begin{aligned}\nabla \cdot \mathbf{J}(\mathbf{e}, u) + \Sigma_t(\mathbf{r}, u)\phi(\mathbf{r}, u) &= \sum_m \int_{u+\ln\alpha_m}^{u} \Sigma_s^m(\mathbf{r}, u')f_{s0}^m(\mathbf{r}, u', u)\phi(\mathbf{r}, u')\,du + S(\mathbf{r}, u)\\ &\quad + Q_0(\mathbf{r}, u)\end{aligned} \tag{7.2}$$

and

$$\frac{1}{3}\nabla\phi(\mathbf{r}, u) + \Sigma_t(\mathbf{r}, u)\mathbf{J}(\mathbf{r}, u) = \sum_m \int_{u+\ln\alpha_m}^{u} \Sigma_s^m(\mathbf{r}, u')f_{s1}^m(\mathbf{r}, u', u)\mathbf{J}(\mathbf{r}, u')\,du' \tag{7.3}$$

where f_{s0}^m and f_{s1}^m refer to the elastic scattering frequencies and $S(\mathbf{r}, u)$ denotes the inelastic scattering source per unit lethargy. Furthermore, the fission source Q is assumed to be isotropic, i.e. $Q_1 = 0$. The inelastic scattering is treated by a formalism equivalent to assuming isotropy in the inelastic scattering kernel $\Sigma_{\text{in}}^m(\mathbf{r}, u', u)$. Therefore

$$S(\mathbf{r}, u) = \sum_m \int_0^u \Sigma_{\text{in}}^m(\mathbf{r}, u', u)\phi(\mathbf{r}, u')\, du'. \tag{7.4}$$

Light moderators and hydrogen in particular exhibit simultaneously large average energy losses per fission and considerable anisotropy of elastic scattering in the laboratory coordinate system. Consequently, elastic scattering by hydrogen is accounted for exactly to the degree allowed by the P_1 truncation error. The hydrogen slowing down integrals η_0 and $\boldsymbol{\eta}_1$ included in eqs. (7.2) and (7.3)

$$\eta_0(u) = \int_0^u \Sigma_s^{\mathrm{H}}(u') f_{s0}^{\mathrm{H}}(u', u)\phi(u')\, du' \tag{7.5}$$

and

$$\boldsymbol{\eta}_1(u) = \int_0^u \Sigma_s^{\mathrm{H}}(u') f_{s1}^{\mathrm{H}}(u', u)\mathbf{J}(u')\, du' \tag{7.6}$$

are treated as new effective variables. Observing that[5]

$$f_{s0}^{\mathrm{H}}(u', u) = \mathrm{e}^{-(u-u')} \tag{7.7}$$

and

$$f_{s1}^{\mathrm{H}}(u', u) = \mathrm{e}^{-3(u-u')/2} \tag{7.8}$$

one is able to derive the differential equations

$$\eta_0 + \frac{\partial \eta_0}{\partial u} = \Sigma_s^{\mathrm{H}}\phi, \tag{7.9}$$

$$\frac{3}{2}\boldsymbol{\eta} + \frac{\partial \boldsymbol{\eta}_1}{\partial u} = \Sigma_s^{\mathrm{H}}\mathbf{J} \tag{7.10}$$

which amend the original P_1 formalism of eqs. (7.2) and (7.3) and thereby provide the necessary number of equations for the system to be well defined, even with the new variables η_0 and $\boldsymbol{\eta}_1$. Due to the simplified spatial dependence in one radial dimension, all the vector quantities, such as $\mathbf{J}$ and $\boldsymbol{\eta}_1$, will be reduced to scalars and it will be implied in the subsequent notation.

Defining the slowing down density $q(\mathbf{r}, u, \mu)$,[6] i.e. the number of neutrons passing the lethargy u per unit volume and time in the direction μ,

$$q(\mathbf{r}, u, \mu) = \int_u^{u-\ln\alpha} du' \int_{u'+\ln\alpha}^{u} du'' \int d\mu' \Sigma_s(\mathbf{r}, u'') f_s(u'', u', \mu_0) \phi(\mathbf{r}, u'') \tag{7.11}$$

where μ' denotes the incident direction, one can infer from a lengthy analysis[6] that η_0 and $\boldsymbol{\eta}_1$ are the first two Legendre components of $q(\mathbf{r}, u)$. This observation facilitates the understanding of the so-called Greuling–Goertzel procedure which is based upon postulation of the relation

$$q_0^m(u) + \lambda^m(u) \frac{\partial q_0^m(u)}{\partial u} = \xi^m(u) \Sigma_s^m(u) \phi(u) \tag{7.12}$$

for any material m other than hydrogen. λ^m is a free parameter to be determined separately[6] and ξ^m denotes the average lethargy gain per collision. Setting $\lambda^m = 0$ reduces the Greuling–Goertzel procedure to the conventional Fermi age theory.[5] A single lumped equation is frequently used to describe the isotropic slowing down from other materials than hydrogen and therefore the index m in eq. (7.12) will be replaced by the notation non-H. Note the similarity between the exact form of eq. (7.9) for hydrogen and the Greuling–Goertzel form of eq. (7.12) for the non-hydrogen elements. The isotropic slowing down density $q_0 = \sum_m q_0^m$ obeys the differential equation

$$\int_{u+\ln\alpha}^{u} \Sigma_s^{\text{non-H}}(u') f_{s0}^{\text{non-H}}(u', u) \phi(u')\, du' = \Sigma_s^{\text{non-H}}(u) \phi(u) - \frac{\partial q_0}{\partial u}. \tag{7.13}$$

Substituting eqs. (7.5), (7.9) and (7.13) into eq. (7.2) yields

$$\nabla \cdot \mathbf{J}(\mathbf{r}, u) + \Sigma_{\text{ne}}(\mathbf{r}, u)\phi(\mathbf{r}, u)$$
$$= -\frac{\partial \eta_0(\mathbf{r}, u)}{\partial u} - \frac{\partial q_0(\mathbf{r}, u)}{\partial u} + S(\mathbf{r}, u) + Q_0(\mathbf{r}, u) \quad (7.14)$$

where the total non-elastic cross-section Σ_{ne} of the system is defined by

$$\Sigma_{\text{ne}} = \Sigma_a + \Sigma_{\text{in}} = \Sigma_t - \Sigma_s^{\text{H}} - \Sigma_s^{\text{non-H}}. \quad (7.15)$$

$\Sigma_s^{\text{non-H}}$ refers to elastic scattering cross-section.

A corresponding reformulation must be carried out for the scattering integral term of eq. (7.3). Hydrogen is treated by means of eqs. (7.6), (7.8) and (7.10), whereas anisotropic scattering in non-hydrogenous material is analysed within the framework of the age theory by introducing the approximation[4]

$$\Sigma_s^m(u')J(u') = \Sigma_s^m(u)J(u). \quad (7.16)$$

Manipulations are parallel to those performed in section 2.3 for the energy variable. Upon inserting eqs. (7.6), (7.10) and (7.16) into eq. (7.3) one obtains

$$\frac{1}{3}\nabla\phi(\mathbf{r}, u) + \Sigma_{\text{tr}}(\mathbf{r}, u)J(\mathbf{r}, u) = -\frac{2}{3}\frac{\partial \eta_1(\mathbf{r}, u)}{\partial u} \quad (7.17)$$

where the transport cross-section Σ_{tr} has been introduced in section 2.3 and the average of the cosine of the scattering angle $\bar{\mu} = 2/3A$ has been used explicitly for hydrogen.

Equations (7.9), (7.10), (7.12), (7.14) and (7.17) incorporate the lethargy transfer approximations used in a series of computer codes[7,8] originating from a programme called MUFT. Note that these considerations pertain to the fast spectrum block of Fig. 7.2. The spatial fast flux dependence is treated in two different degrees of detail, of which the crude approach will be discussed first.

To eliminate the explicit spatial gradient terms the further development will be conducted in a Fourier transformed space by introducing the transformation

$$\hat{g}(B, u) = \int_{-\infty}^{\infty} e^{-iBx} g(x, u)\, dx \tag{7.18}$$

for all the dependent variables $g = \phi, J, \eta_0, \eta_1, q_0, S$ and Q_0. Any given value B of the Fourier transform variable corresponds to a spatial mode $\exp(i|B|x)$ for positive bucklings B^2 and to $\exp(-|B|x)$ for negative B^2.

For physical reasons ϕ and J vanish at infinity and consequently the Fourier transforms of their derivatives obey the formula

$$\frac{\partial \hat{g}}{\partial x} = iB\hat{g} \tag{7.19}$$

which is used in transforming eqs. (7.14) and (7.17). Excluding the introduction of the buckling the lattice cell is considered homogeneous with no dependence on the spatial variable. The Fourier transformed equations obtain now the form

$$iB\hat{J}(B, u) + \Sigma_{ne}(u)\hat{\phi}(B, u)$$

$$= -\frac{\partial \hat{\eta} \cdot (B, u)}{\partial u} - \frac{\partial \hat{q}_0(B, u)}{\partial u} + \hat{S}(B, u) + \hat{Q}_0(B, u), \tag{7.20}$$

$$\frac{1}{3} iB\hat{\phi}(B, u) + \Sigma_{tr}(u)\hat{J}(B, u) = -\frac{2}{3}\frac{\partial \hat{\eta}_1(B, u)}{\partial u}, \tag{7.21}$$

$$\hat{\eta}_0(B, u) + \frac{\partial \hat{\eta}_0(B, u)}{\partial u} = \Sigma_s^H(u)\hat{\phi}(B, U), \tag{7.22}$$

$$\frac{3}{2}\hat{\eta}_1(B, u) + \frac{\partial \hat{\eta}_1(B, u)}{\partial u} = \Sigma_s^H(u)\hat{J}(B, u), \tag{7.23}$$

$$\hat{q}_0(B, u)\lambda \frac{\partial \hat{q}_0(B, u)}{\partial u} = \xi \Sigma_s^{\text{non-H}}(u)\hat{\phi}(B, u). \tag{7.24}$$

To reiterate the essence of the above set, eqs. (7.20)–(7.21) are the P_0 and P_1 equations transformed from eqs. (7.14) and (7.17) and eqs. (7.22)–(7.23) describe isotropic and anisotropic elastic slowing down from hydrogen corresponding to the original relations in eqs. (7.9)–(7.10). Finally, eq. (7.24) accounts for isotropic elastic scattering from non-hydrogenous elements and is obtained from averaging the constants λ^m and $\xi^m \Sigma_s^m$ over all m except $m = \text{H}$.

Prior to proceeding any further, it is important to note that in addition to the P_1 scheme eqs. (7.20)–(7.24) exhibit also a variant procedure designated the B_1 approximation.[6] In the B_N scheme the Fourier transformed transport equation is operated on by $(1 - iB\mu/\Sigma_t)^{-1}$ before inserting the angular Legendre expansions and before proceeding in a manner analogous to the treatment of this section. The B_1 approximation is useful, since a more rapidly converging solution can usually be obtained by the method than is achieved by the P_1 scheme.[6] However, no formal alterations need to be made in the equations and therefore the basis of the B_N method is not discussed here. The reader is rather referred to the literature, e.g. ref. 6. The only change would occur in the transport cross-section appearing in eq. (7.21), where the total cross-section fraction has to be multiplied by a factor γ, i.e. $\Sigma_t \to \gamma\Sigma_t$ with

$$\gamma(u) = \frac{\alpha^2}{3}\frac{\tan^{-1}\alpha}{\alpha - \tan^{-1}\alpha} \tag{7.25}$$

and

$$\alpha(u) = B/\Sigma_t(u). \tag{7.26}$$

The set of eqs. (7.20)–(7.24) involves still complex quantities and dependence on the Fourier transform variable B. The development[7,8] assumes the fundamental mode flux shape

$$\phi(x, u) = f(u)\,e^{iB_0x} \tag{7.27}$$

where the spatial and lethargy variables are implied to be separable. The origin $x = 0$ refers to the centre of the reactor. $f(u)$ is the new dependent variable, whereas B_0 is determined by an approximate leakage consideration. In the crudest case one may choose the buckling B_0^2 to be equal to $B_g^2 = (r_0/R)^2$ where r_0 denotes the lowest zero of the Bessel function J_0. In other words,[5] B_g^2 is obtained as the lowest eigenvalue of the diffusion equation

$$\nabla^2\phi - B_g^2\phi = 0 \tag{7.28}$$

over a bare cylindrical reactor. The fundamental mode approximation made here is adequate for large LWRs with slightly enriched fuel. Fast spectrum and fast cross-section calculation are not very sensitive functions of the buckling approximation. For the current $J(x, u)$ the definition

$$J(x, u) = -D(u)\nabla\phi(x, u) \tag{7.29}$$

is applied resulting in

$$J(x, u) = -i\frac{B_0}{|B_0|}j(u)\,e^{iB_0x} \tag{7.30}$$

with

$$j(u) = |B_0|D(u)f(u). \tag{7.31}$$

Assuming that the isotropic quantities, i.e. S, Q_0, η_0 and q_0, follow the shape of the neutron flux and that η_1 follows the current, eqs. (7.20)–(7.24) are integrated over the whole of the real axis to obtain appropriate equations corresponding to the constant buckling B_0^2. The integrations can be carried out by means of the formula

$$\int_{-\infty}^{\infty} dBg(B)\int_{-\infty}^{\infty} e^{i(B-B_0)x}\,dx = 2\pi g(B_0) \tag{7.32}$$

and one has a new system of equations[7, 8]

$$\frac{B_0^2}{|B_0|}j(u) + \Sigma_{ne}(u)f(u) = -\frac{d\eta(u)}{du} - \frac{dq(u)}{du} + s(u) + r(u), \tag{7.33}$$

$$-\frac{|B_0|}{3}f(u) + \Sigma_{tr}(u)j(u) = -\frac{2}{3}\frac{d\rho(u)}{du}, \tag{7.34}$$

$$\eta(u) + \frac{d\eta(u)}{du} = \Sigma_s^{H}(u)f(u), \tag{7.35}$$

$$\frac{3}{2}\rho(u) + \frac{d\rho(u)}{du} = \Sigma_s^{H}(u)j(u), \tag{7.36}$$

$$q(u) + \lambda\frac{dq(u)}{du} = \xi\Sigma_s^{\text{non-H}}(u)f(u) \tag{7.37}$$

where η, q, s, r and ρ denote the lethargy dependent amplitudes of η_0, q_0, S, Q_0 and η_1, respectively.

Equations (7.33)–(7.37) are discretized in the lethargy variable.

Historically the MUFT scheme employs 54 groups from 10 MeV to 0.625 eV, over which the system of equations is integrated. The treatment of sharp resonances in the cross-sections will be discussed later. At this point the resonance escape probabilities p_n are employed for each lethargy group. Dividing the absorption cross-section $\Sigma_a(u)$ in the resonant and smooth components

$$\Sigma_a(u) = \Sigma^{\text{res}}(u) + \Sigma^{\text{sm}}(u) \tag{7.38}$$

the resonance contribution in the nth group is calculated from

$$\int_n \Sigma^{\text{res}}(u) f(u)\, du = (1 - p_n)(\eta + q)_{n-1/2} \tag{7.39}$$

where $(\eta + q)_{n-1/2}$ is the total slowing down density at the upper energy endpoint of the group. For the smooth cross-sections and for an arbitrary function $g(u)$ the integration is performed by

$$\int_n \Sigma_x(u) g(u)\, du = \Sigma_{x,n} g_n \Delta u_n \tag{7.40}$$

with $\Sigma_{x,n}$ denoting the group cross-sections for the reaction x, g_n being the value of $g(u)$ in the middle of the group interval and Δu_n denoting the group width.

The final step in processing the P_1 or B_1 and Greuling–Goertzel equations is to cast them in a discretized form. The integration of eqs. (7.33)–(7.37) over the lethargy interval Δu_n yields by the application of eqs. (7.39)–(7.40)

$$\frac{B_0^2}{|B_0|} j_n \Delta u_n + \Sigma_{\text{ne},n} f_n \Delta u_n = s_n \Delta u_n + r_n \Delta u_n - \eta_{n+1/2} - q_{n+1/2} + p_n(\eta_{n-1/2} + q_{n-1/2}), \tag{7.41}$$

$$-\frac{|B_0|}{3} f_n \Delta u_n + \Sigma_{\text{tr},n} j_n \Delta u_n = -\frac{2}{3}(\rho_{n+1/2} - \rho_{n-1/2}), \tag{7.42}$$

$$\eta_n \Delta u_n + \eta_{n+1/2} - \eta_{n-1/2} = \Sigma_{sn}^{\text{H}} f_n \Delta u_n, \tag{7.43}$$

$$\frac{3}{2} \rho_n \Delta u_n + \rho_{n+1/2} - \rho_{n-1/2} = \Sigma_{sn}^{\text{H}} j_n \Delta u_n, \tag{7.44}$$

$$q_n \Delta u_n + \lambda_n (q_{n+1/2} - q_{n-1/2}) = \xi \Sigma_{sn}^{\text{non-H}} f_n \Delta u_n \tag{7.45}$$

where the discretized quantities g_n refer to the middle of the nth group while $g_{n-1/2}$ and $g_{n+1/2}$ refer to the endpoints. The inelastic scattering source into the group n is given by

$$r_n \Delta u_n = \sum_j \sum_{m=1}^{n-1} \Sigma_{\text{in},mn}^j f_m \Delta u_m. \tag{7.46}$$

In order to solve eqs. (7.41)–(7.45) for all values of n it is necessary to eliminate the q_n's, η_n's and ρ_n's. These values are combined from those at the endpoints. Since η and ρ describe hydrogen and q_n the rest of the elements it is advisable to use different input weighting factors W_{H} and W_0 as follows:

$$\eta_n = W_{\text{H}} \eta_{n+1/2} + (1 - W_{\text{H}}) \eta_{n-1/2}\,, \tag{7.47}$$

$$\rho_n = W_{\text{H}} \rho_{n+1/2} + (1 - W_{\text{H}}) \rho_{n-1/2}\,, \tag{7.48}$$

$$q_n = W_0 q_{n+1/2} + (1 - W_0) q_{n-1/2}. \tag{7.49}$$

After inserting eqs. (7.47)–(7.49) into eqs. (7.41)–(7.45) the resulting system of linear equations can be solved by elementary techniques. Undoubtedly the values $W_0 = W_{\text{H}} = 0.5$ would lead to convergence, but with rather large group widths and heavy resonance effects somewhat different values of W_{H} and W_0 will yield more acceptable results. The influence of two choices is shown in Fig. 7.4. In view of Fig. 7.4 it is readily obvious that the strong U^{238} resonances induce severe oscillations in a MUFT spectrum calculation for a LWR lattice.

The approach discussed above ignores entirely the lattice heterogeneity. Prior to the homogeneous calculation an infinite lattice of fuel rods is considered in some lattice programme modules and afterwards a flux and volume weighted homogenization is performed. The heterogeneous system is described via escape and collision probabilities.[9] Let P_{ij} denote the probability that a neutron due to a source in the region i will undergo the next collision in the region j. These probabilities are computed for a flat source of neutrons in the fuel (F) and moderator (M) regions and for either $i = F$, $j = M$ or $i = M$, $j = F$. The sources per unit volume are

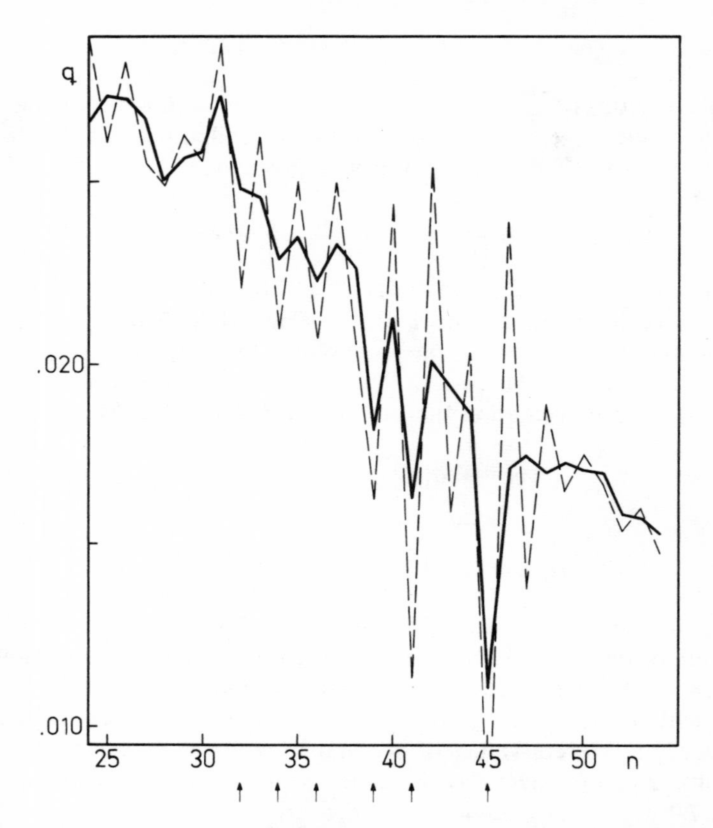

FIG. 7.4. Non-hydrogen slowing down density in LWR lattice.[8] The solid line corresponds to weighting $W_H = 0.55$, $W_0 = 0.90$, while the dotted one is for $W_H = W_0 = 0.50$.

denoted by Q_i with either $i = F$ or $i = M$. The transport of neutrons is governed now by the equations

$$\Sigma_t^F(u)\phi_F(u) = [1 - P_{FM}(u)]Q_F(u) + \gamma P_{MF}(u)Q_M(u)\,, \quad (7.50)$$

$$\gamma\Sigma_t^M(u)\phi_M(u) = P_{FM}(u)Q_F(u) + \gamma[1 - P_{MF}(u)]Q_M(u) \quad (7.51)$$

where Σ_t^F and Σ_t^M are the total cross-sections of fuel and moderator, respectively, while γ denotes the moderator-to-fuel volume ratio. Self-explanatorily, $\phi_F(u)$ and $\phi_M(u)$ denote the average neutron fluxes per unit lethargy in fuel and moderator.

The neutron transfer eqs. (7.50)–(7.51) are augmented by the description of the elastic slowing down, i.e. by

$$\eta^i(u) + \frac{d\eta^i(u)}{du} = \Sigma_s^{H,i}(u)\phi_i(u) \tag{7.52}$$

for hydrogen and by

$$q^i(u) + \lambda^i \frac{dq^i(u)}{du} = \xi\Sigma_s^{\text{non-H},i}(u)\phi_i(u) \tag{7.53}$$

for the elastic slowing down in non-hydrogenous media. Equations (7.52)–(7.53) are essentially the same as eqs. (7.35) and (7.37). The index i obtains either $i = F$ for fuel and $i = M$ for moderator. The source terms of eqs. (7.50)–(7.51) comprise fission, as well as elastic and inelastic scattering

$$Q_i = Q_i^f + Q_i^{el} + Q_i^{in} \tag{7.54}$$

where the fission source

$$Q_i^f = (1+\gamma)r(u)\delta_{iF} \tag{7.55}$$

is nonzero in fuel only (δ_{ij} is the Kronecker delta). The elastic term is defined by

$$Q_i^{el} = \Sigma_s^i(u)\phi_i(u) - \frac{d\eta^i(u)}{du} - \frac{dq^i(u)}{du} \tag{7.56}$$

and the inelastic source becomes simply

$$Q_i^{in} = \sum_j \int_0^u \Sigma_{in}^j(u', u)\phi_i(u')\,du' \tag{7.57}$$

where j goes over all the elements in the region i.

As soon as the transfer probabilities P_{FM} and P_{MF} are known, eqs. (7.50)–(7.53) can be solved by elementary discretization techniques. The determination of the probability functions is principally described in ref. 9. The presence of an infinite array of fuel rods must be taken into account here and numerous approximate procedures are devised in the literature for making this Dancoff correction. Introducing the escape probabilities P_i ($i = F$ or $i = M$)

$$P_i(u) = \frac{B^i(u)}{\Sigma_t^i(u)l_i} \qquad (7.58)$$

by means of the auxiliary functions $B^j(u)$, fairly good approximations of $b^i(u)$ are given[10, 11]

$$\beta^F(u) = 1 - [1 + \Sigma_t^F(u)l_F/(n+1)]^{-(n+1)} \qquad (7.59)$$

and

$$\beta^M(u) = 1 - \frac{\exp[-\tau\Sigma_t^M(u)l_M]}{1 + (1-\tau)\Sigma_t^M(u)l_M} \qquad (7.60)$$

where l_F denotes the average chord length of fuel rods and is equal to the rod diameter, $l_M = \gamma l_F$, $n = 3.58$ for cylindrical rods[10] and τ is a geometry-dependent rod separation parameter

$$\tau = \left(\frac{1}{\sqrt{\frac{\pi}{4\alpha}(1+\gamma)}} - 1\right) \Big/ \gamma - \tau \qquad (7.61)$$

with the values $\alpha = 1$, $\tau = 0.08$ for square lattices and $\alpha = \sqrt{3/2}$, $\tau = 0.12$ for hexagonal ones. Equation (7.61) is valid for most practical rod arrays, but certain improvements may still be made if desired. Note that $\beta'(u)$'s are also proportional to the characteristic probability for a neutron entering the region i to have a collision there after a number of previous traversals of fuel. $\beta^M(u)$ is also identical to the Dancoff factor and $1 - \beta^M$ is the Dancoff correction.[9] The probabilities P_{FM} and P_{MF} are now obtained from

$$P_{ij}(u) = \frac{P_i(u)\beta^j(u)}{1 - [1 - \beta^i(u)][1 - \beta^j(u)]} \qquad (7.62)$$

with either $i = F$ and $j = M$, or conversely.

$P_{ij}(u)$ obeys the fundamental reciprocity relation[9]

$$V_i \Sigma_t^i(u) P_{ij}(u) = V_j \Sigma_t^j(u) P_{ji}(u) \tag{7.63}$$

which can be used here and which is vital in the resonance calculation to be discussed later.

Equations (7.50)–(7.57) can now be discretized in analogy with the approach applied to the truly homogeneous systems considered first in this section. The resulting equations have the form:

$$\Sigma_{tn}^F \phi_{Fn} = (1 - P_{FMn}) Q_{Fn} Q_{Mn}, \tag{7.64}$$

$$\gamma \Sigma_{tn}^M \phi_{Mn} = P_{FM} 1_{Fn} + \gamma (1 - P_{MFn}) Q_{Mn}, \tag{7.65}$$

$$\eta_n^i \Delta u_n + \eta_{n+1/2}^i - \eta_{n-1/2}^i = \Sigma_{sn}^{\mathrm{H},i} \phi_{\mathrm{in}} \Delta u_n, \tag{7.66}$$

$$q_n^i \Delta u_n + \lambda_n^i (q_{n+1/2}^i - q_{n-1/2}^i) = \xi \Sigma_{sn}^{\text{non-H},i} \phi_{\mathrm{in}} \Delta u_n, \tag{7.67}$$

$$Q_{\mathrm{in}} = Q_{\mathrm{in}}^f + Q_{\mathrm{in}}^{\mathrm{el}} + Q_{\mathrm{in}}^{\mathrm{in}}, \tag{7.68}$$

$$Q_{\mathrm{in}}^f = (1 + \gamma) r_n \delta_{iF}, \tag{7.69}$$

$$Q_{\mathrm{in}}^{\mathrm{el}} = \Sigma_{sn}^i \phi_{\mathrm{in}} - (\eta_{n+1/2}^i - \eta_{n-1/2}^i + q_{n+1/2}^i - q_{n-1/2}^i)/\Delta u_n, \tag{7.70}$$

$$Q_{\mathrm{in}}^{\mathrm{in}} = \sum_{m=1}^{n-1} \Sigma_{\mathrm{in},mn}^i \phi_{im} \Delta u_m / \Delta u_n \tag{7.71}$$

where $i = F, M$ and n is the group index.

The linear system of eqs. (7.64)–(7.71) can be solved by numerical algorithms. The main output from the fast spectrum calculations consists of the few group cross-sections to which the original 54 groups are condensed by flux weighting procedures. The weighting fluxes are obtained from either the homogeneous calculation eqs. (7.41)–(7.45), or the two region Dancoff calculations based on eqs. (7.64)–(7.71). To demonstrate the magnitude of the fast advantage factor ϕ_F/ϕ_M over the first ten MUFT energy groups 0.82–10 MeV, comparative results are summarized in Table 7.4 for the heterogeneous method described above. The reference is a more accurate Monte Carlo calculation.[12]

TABLE 7.4
Fast Advantage Factor ϕ_F/ϕ_M *within the Range* 0.82–10 MeV

γ	Method	
H_2O/U*	MC[12]	FORM[8]
1	1.09	1.095
4	1.31	1.356

*Rod diameter 1.5 cm.

The actual group collapsing is accomplished by taking a weighted sum of the parameters over the new group lethargy intervals. If m is the few group index, then the condensed group source, flux and current are defined by

$$Q_m = \sum_n r_n \Delta u_n , \tag{7.72}$$

$$\phi_m = \sum_n f_n \Delta u_n , \tag{7.73}$$

$$J_m = \sum_n j_n \Delta u_n \tag{7.74}$$

where the sum extends over all the fine groups n within the coarser interval m.

The rates for the reaction x to occur are sums over the multigroup reaction rates

$$R_{xm} = \sum_n \Sigma_{xn} f_n \Delta u_n \tag{7.75}$$

from which the few group cross-sections are readily obtained by

$$\Sigma_{xm} = R_{xm}/\phi_m. \tag{7.76}$$

Note also that the total leakage L_n in group n is given by

$$L_n = \frac{B_0^2}{|B_0|} j_n \Delta u_n \tag{7.77}$$

[cf. the first term in eq. (7.41)] and coarse group leakage follows the direct rule of summation.

The diffusion constant D has already been defined in eq. (7.31) and the fine group structure implies a similar relation. In the few group notation

$$D_m = \frac{J_m}{|B_0|\phi_m}. \tag{7.78}$$

It is clear that a number of other few group quantities can be edited in the output if desired. All procedures are closely related to the ones in eqs. (7.72)–(7.78) where the most important parameters are shown.

For a breeder the spectrum calculation comprises naturally the fast energy range only. The P_N or B_N procedures used there and sometimes even for thermal reactors have been developed in a less heuristic direction; the equations are discretized directly from the Fourier transformed form. An outline will be given in section 8.6.

Resonance Calculation

The physical basis of resonance phenomena was discussed in section 4.2 in connection with the fuel temperature feedback. Although the typical resonance energies lie within the MUFT domain a separate and more careful analysis must be conducted as indicated by a distinct box in Fig. 7.2. The treatment aims to provide the resonance escape probabilities appearing in eqs. (7.39) and (7.41).

Apart from the complicated resonance structure of the U^{238} absorption cross-section, the neutron flux is of course determined by the moderator cross-sections. The microscopic scattering cross-section of hydrogen is shown in Fig. 7.5, exhibiting remarkably constant values over the main resonance region below 10 keV down to the lower end of the MUFT energy domain.

In the first approximation of Fermi age theory, eq. (7.12) is assumed to be valid in the slowing down process for the entire lattice with

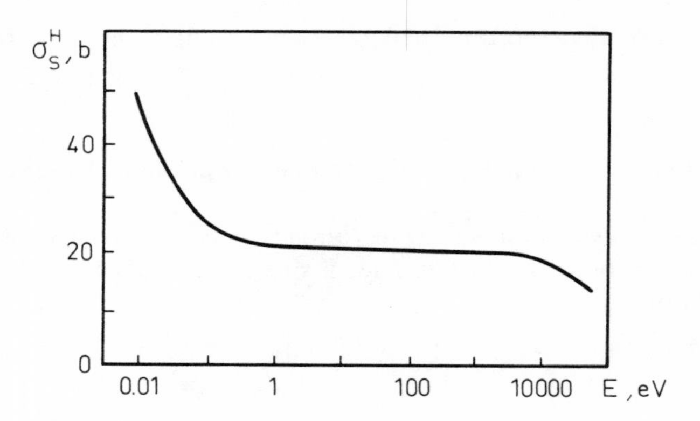

FIG. 7.5. Microscopic scattering cross-section for hydrogen.

$\lambda(dq_0/du) = 0$ (cf. ref. 5) and one has for the slowing down density

$$q(u) = \xi\Sigma_s(u)\phi(u) \tag{7.79}$$

where all the sub- and superscripts are omitted for convenience. Based on eq. (7.79), further insight into the neutron behaviour can be gained by ignoring the absorption which in any case is substantial only in the vicinity of resonances. With no leakage, i.e. infinite medium $B_0^2 = 0$, the slowing down source must be constant and consequently[5]

$$\phi(u) = \frac{1}{\xi\Sigma_s(u)} \tag{7.80}$$

where the source strength is normalized to unity. Because of the stringent assumptions eq. (7.80) can be expected to hold only in asymptotic conditions, i.e. for a neutron having already experienced a number of collisions. In view of the hydrogen scattering cross-section of Fig. 7.5 it is readily understandable that the flux $\phi(u)$ in Fig. 1.10 obtains an almost constant value over a considerable energy range.

To study the different resonance approximations one states the zero-buckling transport equation

$$\Sigma_t(u)\phi(u) = \int_{u+\ln\alpha_A}^{u} \Sigma_s^A(u', u)\phi(u')\,du' + \sum_j \int_{u+\ln\alpha_j}^{u} \Sigma^j(u', u)\phi(u')\,du' \quad (7.81)$$

where the principal absorber element denoted by A is taken separately and the sum extends over the nuclear species present in the uniform homogeneous mixture considered. The elastic scattering kernel for isotopic scattering in the centre-of-mass system obeys within the range of integration the formula[5]

$$\Sigma_s(u', u) = \Sigma_s(u)\,e^{-(u-u')}/(1-\alpha). \quad (7.82)$$

In the narrow resonance (NR) approximation, which applies when the resonance widths are small in comparison with the energy loss $(1-\alpha)E_r$, the integral is evaluated as if the contribution over the resonance were negligible. The scattering cross-section is replaced by the potential scattering term, i.e. the last term in eq. (4.6). For the flux the asymptotic constant value is used. A typical integral in eq. (7.81) now becomes

$$\int_{u+\ln\alpha}^{u} \Sigma_s(u')\phi(u')\,e^{-(u-u')}/(1-\alpha)\,du' = \Sigma_p(u) \quad (7.83)$$

where the asymptotic flux is normalized to unity and Σ_p denotes the potential scattering cross-section.

At the other extreme of wide resonances (WR) the energy loss or lethargy increment per collision is small in comparison with the resonance width and it is then consistent to assume a constant flux and scattering cross-section over the range of integration. One obtains therefore

$$\int_{u+\ln\alpha}^{u} \Sigma_s(u')\phi(u')\,e^{-(u-u')}/(1-\alpha)\,du' = \Sigma_s(u)\phi(u). \quad (7.84)$$

Note that the WR approximation is applicable to heavy fuel isotopes only, whereas the light moderators with sizeable lethargy gains per collisions are treated in the NR approximation. In the NR and WR approximations eq. (7.81) yields for the flux $\phi(u)$

$$\phi_{NR}(u)=\frac{\Sigma_p^A+\sum_j \Sigma_s^j}{\Sigma_a^A(u)+\Sigma_s^A(u)+\sum_j \Sigma_s^j} \tag{7.85}$$

and

$$\phi_{WR}(u)=\frac{\sum_j \Sigma_s^j}{\Sigma_a^A(u)+\sum_j \Sigma_s^j} \tag{7.86}$$

respectively. In eqs. (7.85) and (7.86) an implicit assumption is made that the absorption part of the total cross-section is due to the main absorber only.

More sophisticated methods known as intermediate resonance (IR) approximations, e.g. ref. 12, introduce a free parameter λ to facilitate an expression intermediate to eqs. (7.85)–(7.86). Employing microscopic cross-sections a possible intermediate flux expression is

$$\phi_{IR}(u)=\frac{\lambda\sigma_P^A+\sum_j \sigma_s^j}{\sigma_a^A(u)+\lambda\sigma_s^A(u)+\sum_j \sigma_s^j} \tag{7.87}$$

which reduces to NR and WR approximations if λ is selected properly, viz.

$$\begin{aligned}\lambda &= 1, \text{NR},\\ \lambda &= 0, \text{WR}.\end{aligned} \tag{7.88}$$

σ_s^j is calculated per one absorber atom

$$\sigma_s^j=\Sigma_s^j/N^A \tag{7.89}$$

and actually is not the proper elastic microscopic scattering cross-section.

Among the methods used for calculating the value λ, the one due to Ishiguro[12] is outlined below. In this method λ is determined from the requirement of obtaining the best possible value for the resonance integral defined in eq. (4.18). Introducing x from eq. (4.7) as a new variable of integration the exact resonance integral can be written as

$$RI_i^r = \frac{\Gamma}{2E_r}\int_{-\infty}^{\infty} \sigma_i(x)\phi_{IR}(x)\left[1+\frac{\phi(x)-\phi_{IR}(x)}{\phi_{IR}(x)}\right] dx \qquad (7.90)$$

with the minor inaccuracy of extending the range of integration. In eq. (7.90) $\phi(x)$ denotes the exact flux. Clearly the IR approximation corresponds to the exact treatment as soon as the second term in the brackets vanishes, i.e.

$$\int_{-\infty}^{\infty} \sigma_i(x)[\phi_{IR}(x)-\phi(x)]\, dx = 0. \qquad (7.91)$$

Substituting the flux ϕ_{IR} from (7.87) into eq. (7.91) λ is determined from[12]

$$\lambda = 1 - \frac{\int_{-\infty}^{\infty} \sigma_i(x)\phi_{IR}(x)[\sigma_P^A - K_A(\sigma_s^A\phi)]\, dx}{\int_{-\infty}^{\infty} \sigma_i(x)\phi_{IR}(x)[\sigma_P^A - \sigma_s^A(x)\phi(x)]\, dx} \qquad (7.92)$$

where K_i denotes the operator

$$K_i(\sigma\phi) = \int_{u+\ln\alpha_i}^{u} \sigma(u')\phi(u')\, e^{-(u-u')}/(1-\alpha_i)\, du'. \qquad (7.93)$$

Naturally $\phi(x)$ is not known, but it must be approximated by suitable trial functions or the procedure must be made iterative.

The above outline of a homogeneous resonance calculation lends itself readily to the extensions for regarding temperature-dependent effects, interaction of closely spaced resonances with each other and for the analysis of the unresolved resonance region where no distinct resonance energies are established. Many of these aspects have been incorporated in the computer codes available.[13] A more important aspect is the consideration of spatial heterogeneities to which the resonance neutrons are subject.

The discussion on the cladding contribution to the resonance calculation will be conducted later, and the analysis is commenced by introducing the coupled neutron transfer equations for the fuel and moderator regions. Two different types of moderator will be considered besides the main absorber A. There is allowed to be one type of moderator material admixed homogeneously with the fuel in the volume V_F, whereas the proper moderator occupies the moderator volume V_M. The scattering cross-sections of these moderator elements will be denoted by Σ_m^j and Σ_e^j depending on whether the material j is mixed (m) with or external (e) to the fuel absorber A. The neutron transfer is basically governed by two equations of the type of eqs. (7.50)–(7.51) with proper regard to the elastic scattering kernels which now contribute the source terms. The equations are

$$\Sigma_t^F \phi_F = \left[K_A(\Sigma_s^A \phi_F) + \sum_j K_{mi}(\Sigma_m^i \phi_F)\right] P_{FF} + \gamma \sum_j K_{ej}(\Sigma_e^j \phi_M) P_{MF}, \tag{7.94}$$

$$\gamma \Sigma_t^M \phi_M = \left[K_A\left(\Sigma_s^A \phi_F\right) + \sum_j K_{mi}(\Sigma_m^i \phi_F)\right] P_{FM} + \gamma \sum_j J_{ej}(\Sigma_e^j \phi_M) P_{MM} \tag{7.95}$$

where the K operators are defined in eq. (7.93), γ denotes the ratio V_M/V_F and the collision probabilities P_{ij} are introduced earlier in this section.

Equations (7.94)–(7.95) are divided by the absorber number density N^A; in other words, the development will be carried further in terms of the microscopic cross-sections. As in eq. (7.89) σ_t^F and σ_m^i will now be calculated per one absorber atom. For the external moderating elements j a quantity s_j is defined by

$$s_j = \frac{\gamma \Sigma_e^j}{N^A}. \tag{7.96}$$

Equations (7.94)–(7.95) are rewritten

$$\sigma_t^F \phi_F = \left[K_A(\sigma_s^A \phi_F) + \sum_i \sigma_m^i K_{mi}(\phi_F)\right](1 - P_{FM}) + \sum_j s_j K_{ej}(\phi_M) P_{MF}, \tag{7.97}$$

$$s^e \phi_M = \left[K_A(\sigma_s^A \phi_F) + \sum_i \sigma_m^i K_{mi}(\phi_F)\right] P_{FM} + \sum_j s_j K_{ej}(\phi_M)(1 - P_{MF}) \tag{7.98}$$

with s^e standing for

$$s^e = \sum_j s_j. \tag{7.99}$$

Since the cladding is ignored in eqs. (7.94)–(7.98) P_{FM} and P_{MF} obey the reciprocity relation in the form presented by eq. (7.63) which in the present notation can be expressed by

$$\sigma_t^F P_{FM} = s^e P_{MF}. \tag{7.100}$$

Again, as in the case of spectrum calculation, the self- and mutual shielding of the fuel rods is taken into account by the Dancoff correction and a possible approximation of P_{FM} is of the form[9]

$$P_{FM} = \frac{s}{s + \sigma_t^F} \tag{7.101}$$

where s is defined by

$$s = \frac{a(1-C)}{N^F l_F}. \tag{7.102}$$

In this instance it is customary to denote the Dancoff correction by C instead of $\beta^M = 1 - C$. A useful approximation for β^M appears in eq. (7.60). In eq. (7.102) a denotes a further correction expressed often by

$$a = \frac{a_1}{1 + a_2 C}. \tag{7.103}$$

There are a number of suggestions[14] concerning the values of a_1 and a_2.

Substitution of eqs. (7.100)–(7.101) into eqs. (7.97)–(7.98) yields

$$(\sigma_t^F + s)\phi_F - K_A(\sigma_s^A \phi_F) - \sum_i \sigma_m^i K_{mi}(\phi_F) = \frac{s}{s^e} \sum_j s_j K_{ej}(\phi_M) \tag{7.104}$$

and

$$K_A(\sigma_s^A \phi_F) + \sum_i \sigma_m^i K_{mi}(\phi_F)$$

$$= \frac{s^e}{s}(\sigma_t^F + s)\phi_M - \frac{1}{s}\left(\sigma_t^F + s - \sigma_t^F \frac{s}{s^e}\right) \sum_j s_j K_{ej}(\phi_M). \tag{7.105}$$

The IR approximation is introduced in eqs. (7.104)–(7.105).

A way of expressing the IR assumption in eq. (7.87) is to specify a relation for the K_A operator. If λ is again the IR parameter for the absorber nucleus, then the IR approximation is tantamount to

$$K_A(\sigma_s^A \phi_F) = \lambda \sigma_p^A + (1-\lambda)\sigma_s^A \phi_F. \tag{7.106}$$

Introducing corresponding IR parameters for the admixed and for the external moderator and denoting them by κ and η, respectively, the analogous IR equations for the two kinds of moderator nuclei are of the form

$$K_{mi}(\phi_F) = \kappa_i + (1-\kappa_i)\phi_F \tag{7.107}$$

and

$$K_{ej}(\phi_M) = \eta_j + (1-\eta_j)\phi_M \tag{7.108}$$

where i and j go over all the moderating nuclear species present in the lattice cell.

The development proceeds in a manner similar to the treatment of the homogeneous mixture. The resonance parameters are solved from the requirement of eq. (7.91). The flux estimate and the resonance parameters may be improved iteratively.

The cladding can be included in the analysis by a number of methods. First, the cladding can be homogenized into the moderator by defining homogenized constants Σ_{hom} from

$$\Sigma_{\text{hom}} = \frac{V_C \Sigma_t^C + V_M \Sigma_t^M}{V_C + V_M} \tag{7.109}$$

where the indices C and M refer to the cladding and moderator, respectively. An equally well-justified procedure is to surround the fuel rod by a vacuum region.[11] The zircaloy alloys used as cladding in many thermal reactors interact minimally with neutrons and this method can be successful. A further possibility is to divide the cladding into two annular shells, an inner vacuum region and an exterior region, where the group constants are those of the moderator. The division should be performed in a manner which leaves both the physical and optical thicknesses of the cladding unchanged.[13]

The utilization of IR methods in resonance calculations has

become justified after rapid computer routines have evolved. In many cases the IR schemes can be used as an option and more straightforward approaches can be used. A formerly popular method has been the NRIM scheme where the NR approximation is used for the moderator, whereas the heavy absorber A is treated in the WR frame and in the infinite mass (IM) limit $\alpha_A \rightarrow 1$. Lucid descriptions of the method appear in refs. 5 and 9.

Returning to the calculations of the resonance integrals RI, the Breit–Wigner formulae of eqs. (4.12)–(4.16) are employed in eq. (7.87) and one obtains

$$\phi_{\mathrm{IR}}(x) = \frac{b}{\psi(x, \theta) + a\sigma(x, \theta) + b} \tag{7.110}$$

where

$$a = \frac{\lambda \Gamma \sigma_{op}}{\sigma_r(\Gamma_\gamma + \Gamma_f + \lambda \Gamma_n)} \tag{7.111}$$

and

$$b = \frac{\Gamma\left(\lambda\sigma_p + \sum_j \sigma_s^j\right)}{\sigma_r(\Gamma_\gamma + \Gamma_f + \lambda \Gamma_n)}\,. \tag{7.112}$$

σ_{op} denotes the quantity $\sigma_r(R/\lambda\!\!\!^{-})$ appearing in eq. (4.6). The pertinent expression of the resonance integral can be cast into the form[9]

$$RI_i = \frac{2I_{o,i}}{\pi} bJ(\theta, a, b) \tag{7.113}$$

with $i = a$, γ or f,

$$I_{o,i} = \frac{\pi \sigma_r \Gamma_i}{2E_r} \tag{7.114}$$

and

$$J(\theta, a, b) = \frac{1}{2}\int_{-\infty}^{\infty} \frac{\psi(x, \theta)}{\psi(x, \theta) + a\chi(x, \theta) + b}\, dx\,. \tag{7.115}$$

In a crude approximation with $T = 0°\mathrm{K}$ and no interference scattering the resonance integral becomes $RI_i = I_{o,i}/\gamma$ where

$$\gamma = \sqrt{\frac{b^2 + b - a^2}{b^2}}. \tag{7.116}$$

The resonance calculation is linked to the fuel lattice module by means of the resonance escape probabilities p_n appearing in eqs. (7.39) and (7.41). Let u_1 and u_2 denote the limits of the lethargy interval concerned and let $q(u)$ be the slowing down density. The resonance escape probability from u_1 and u_2 is defined by

$$p = \frac{q(u_2)}{q(u_1)} = 1 - \frac{q(u_1) - q(u_2)}{q(u_1)}$$

$$= 1 - \frac{1}{q(u_1)} \int_{u_1}^{u_2} \Sigma_a(u)\phi(u)\, du. \tag{7.117}$$

Approximately p is obtained from

$$p = \exp\left(-\frac{1}{q(u_1)} \int_{u_1}^{u_2} \Sigma_a(u)\phi(u)\, du\right) \tag{7.118}$$

where p_n corresponds to the group limits $u_1 = u_{n-1/2}$, $u_2 = u_{n+1/2}$.

The slowing down density $q(u_1)$ can be chosen from the approximate form of eq. (7.79). Note that Σ_p^A can be used instead of Σ_s^A for the absorber. If there are both admixed and external moderating nuclei present, then eq. (7.118) is written out as

$$p^r = \exp\left[-\frac{RI_a^r}{\xi_A \sigma_p^A + \sum_i \xi_i \sigma_m^i + \sum_j \xi_j s_j}\right] \tag{7.119}$$

for a resonance at E_r. p_n is the product of all p^r with E_r lying within the nth group,

$$p_n = \Pi r^r$$
$$r \in \Delta u_{ji}.$$

For the sake of completeness the condensation of the resonance escape probabilities will be briefly covered here. Few group removal cross-sections Σ_{rm} are calculated, assuming neutron transfer only to the next upper lethargy group. For $m = 1$ the expression has the form

$$\Sigma_{r1} = \frac{Q_i}{\phi_1} - B_0^2 D_1 - \Sigma_{a1} \tag{7.120}$$

whereas for $m > 1$

$$\Sigma_{rm} = \frac{Q_m + \Sigma_{rm-1}\phi_{m-1}}{\phi_m} - B_0^2 D_m - \Sigma_{am} \tag{7.121}$$

where the group parameters are defined in eqs. (7.72)–(7.78). The few group absorption escape probability would then be obtained from

$$p_m = \frac{\Sigma_{rm}}{\Sigma_{rm} + \Sigma_{am}}. \tag{7.122}$$

Thermal Spectrum

Earlier in this section the neutron slowing down from the fission energies has been studied in a fuel pin cell. Also the resonance calculation was actually carried out to complete and to amend the pin cell analysis within the fast energy domain. Further discussion will now be focused on the thermal energy range in the pin cell block of Fig. 7.2.

Lattice geometry will be considered in greater detail than was done in the slowing down calculation. This is in order to recognize the shorter diffusion length of thermal neutrons whose behaviour becomes more sensitive to the local heterogeneities and the presence of the separate fuel, cladding and moderator regions. The methods used in the reactivity control as well as plutonium buildup tend further to distort the thermal flux. In particular, it is worth while to observe that the 0.3 eV resonance of Pu^{239} and the 1.0 eV resonance of Pu^{240} are included in the thermal rather than being treated in the resonance mode.

In a discussion of computer methods one cannot ignore the classic thermalization code THERMOS[(15)] based upon the integral transport formalism of eqs. (2.72)–(2.73). Despite some numerical deficiencies in the original version, the THERMOS method of treating the fuel lattice cell has been widely accepted and many of the subsequent improvements rely heavily on the experience gained by THERMOS. Later in the development some of the more recent and efficient approaches will be discussed.

Confining eq. (2.72) into the thermal energy range it is rewritten as

$$\phi(\mathbf{r}, E) = \int_{V'} K(\mathbf{r}, \mathbf{r}', E)\left\{\int_0^{E_{\rm th}} \Sigma_s(\mathbf{r}', E', E)\phi(\mathbf{r}', E')\, dE' + Q_0(\mathbf{r}', E)\right\} d\mathbf{r}' \tag{7.123}$$

where E_{th} is the thermal cutoff energy and $Q_0(\mathbf{r}', E)$ is the thermalization source

$$Q_0(\mathbf{r}, E) = \int_{E_{th}}^{\infty} \Sigma_s(\mathbf{r}, E', E)\phi(\mathbf{r}, E')\, dE'. \qquad (7.124)$$

The spatial discretization of eq. (7.123) will be executed here in more general terms than subsequent utilization would strictly necessitate.

The collision probability method[9] is traditionally used for the solution of eq. (7.123). Dividing the lattice volume V into separate regions V_i, $1 \le i \le I$, eq. (7.123) is multiplied by $\Sigma_t(\mathbf{r}, E)$ and integrated over a given region V_j. Both the flux and the source including the cross-sections are assumed to be constant within each volume element. Performing parallel by the division of the energy variable into energy groups one gets the discretized equation

$$V_j \Sigma_{tn}^j \phi_{jn} = \sum_{i=1}^{I} V_i P_{ijn} \left[\sum_{m=1}^{N} \Sigma_{smn}^i \phi_{im} + q_n^i \right] \qquad (7.125)$$

where

$$P_{ijn} = \frac{1}{V_i \Delta E_n} \int_{V_j} d\mathbf{r}' \int_{\Delta E_n} dE \frac{\Sigma_t(\mathbf{r}, E)}{4\pi |\mathbf{r} - \mathbf{r}'|^2} \mathrm{e}^{-\alpha(\mathbf{r}', \mathbf{r}, E)} \qquad (7.126)$$

represents the probability for a neutron in the group n born in V_i to undergo its first collision in V_j. The quantities Σ_{tn}^i, ϕ_{in}, Σ_{smn}^i and q_n^i are the constant values of the group parameters Σ_{tn}, ϕ_n, Σ_{smn} and q_n [cf. eqs. (2.40)–(2.45)] specified or to be solved in V_i. Equivalently, if $y(\mathbf{r})$ denotes any of these functions, y^i would be obtained as a spatial average

$$y^i = \frac{1}{V_i} \int_{V_i} y(\mathbf{r})\, d\mathbf{r}. \qquad (7.127)$$

In the solution of eq. (7.125) there are two separate tasks: one concerning the computation of the collision probabilities P_{ijn}, the other the numerical solution of the linear system. Since the numerical part is analogous to those appearing in other subsequent phases, it will be discussed in section 7.6. Formally, one defines vectors $\boldsymbol{\phi}$ and $\mathbf{q}$ with the components ϕ_{in} and q_n^i, respectively, and matrices $\mathbf{P}$ and Σ_s with elements

$$\frac{V_i P_{ijn}}{V_j \Sigma_{tn}^d} \quad \text{and} \quad \Sigma_{smn}^i.$$

Equation (7.125) can then be written in a matrix form simultaneously for all $i \leqslant I$ and $n \leq N$:

$$\boldsymbol{\phi} = \mathbf{P}(\Sigma_s \boldsymbol{\phi} + \mathbf{q}) \tag{7.128}$$

where the iteration techniques are applied.

The calculation of the P_{ijn}'s is based on the spatial mesh spanned over the pin cell. The fine group calculations within the thermal energies are performed in one-dimensional cylindrical geometry and therefore it is relevant to obtain the collision probabilities between different annular mesh regions V_i. In terms of the energy variable, P_{ijn}'s are determined for monoenergetic neutrons corresponding to the group ΔE_n. Suppose first that there is a unit line source along the axis of a cylinder. The probability for a neutron to leak without a collision from a cylinder whose radius is r corresponding to the optical thickness of $\alpha = \Sigma_t r$ can be found from the monoenergetic variant of eq. (7.126) by setting V_i equal to the cylindrical volume and V_j equal to the rest of space. The referred probability is $Ki_2(\alpha)$ where the Bickley functions, e.g. ref. 16, are defined by

$$Ki_n(x) = \int_0^{\pi/2} \sin^{n-1}\theta \exp\left(-\frac{x}{\sin\theta}\right) d\theta. \tag{7.129}$$

To proceed further, consider the discretized annular cell structure depicted in Fig. 7.6. A neutron traversing from A in V_i to B in V_j can enter either once or twice the volume V_j depending on the path direction. Denoting the mesh radii by r_i such that the volume V_i is confined by r_{i-1} and r_i and letting y be fixed, then t is used to express the projection of the radius drawn for the given value y as shown in Fig. 7.6. One is now able to deduce the probability for a neutron emitted from point A to undergo a collision in V_j based on the non-collision probability $Ki_2(\alpha)$. If the distance x' from the surface r_i to A varies over all (t_i, t_{i-1}) corresponding to a constant source in V_i, then the region i to region j collision probability $P_{ij}(y)$ be-

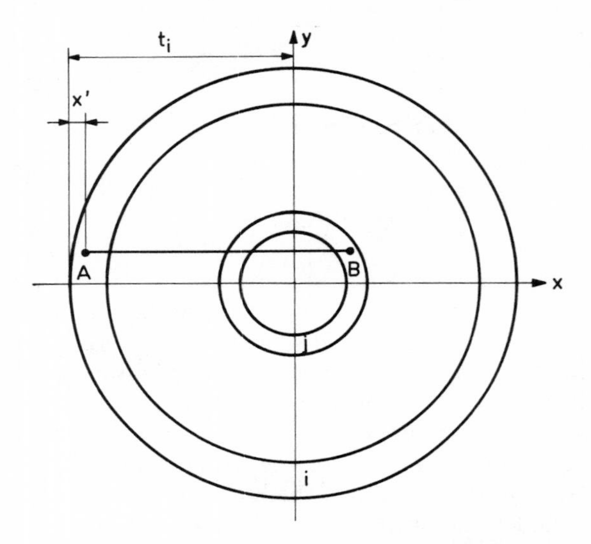

FIG. 7.6. Cell geometry in the collision probability method.

comes[17] for a fixed y

$$
\begin{aligned}
P_{ij}(y) = \frac{1}{t_i - t_{i-1}} \int_0^{t_i - t_{i-1}} & \{Ki_2[\Sigma_t^i(t_i - t_{i-1} - x') + (\tau_{i-1} - \tau_j)] \\
& - Ki_s[\Sigma_t^i(t_i - t_{i-1} - x') + (\tau_{i-1} - \tau_{j-1})] \\
& + Ki_s[\Sigma_t^i(t_i - t_{i-1} - x') + (\tau_{i-1} + \tau_{j-1})] \\
& - Ki_2[\Sigma_t^i(t_i - t_{i-1} - x') + (\tau_{i-1}\tau_j)]\}\, dx' \qquad (7.130)
\end{aligned}
$$

where τ denotes the optical thickness obtained by projection from α similar to that by which t was obtained from r. Note that the first two terms in eq. (1.30) describe the neutrons colliding after having entered for their first time in V_j, while the last two terms give the probability to undergo a collision after the second entrance. Performing the integration in eq. (7.130) yields[16, 17]

$$
P_{ij}(y) = \frac{1}{\Sigma_t^i(t_i - t_{i-1})} \{[Ki_3(\tau_i + \tau_j) - Ki_3(\tau_i - \tau_j)]
$$

$$- [Ki_3(\tau_i + \tau_{j-1}) - Ki_3(\tau_i - \tau_{j-1})]$$
$$- [Ki_3(\tau_{i-1} + \tau_j) - Ki_3(\tau_{i-1} - \tau_j)]$$
$$+ [Ki_3(\tau_{i-1} + \tau_{j-1}) - Ki_3(\tau_{i-1} - \tau_{j-1})]\}. \qquad (7.131)$$

The final step of calculating P_{ijn} for a given energy group consists of integrating eq. (7.132) over y from zero to r_i or r_j depending on whichever is smaller. In practice the collision probabilities are computed numerically[16] rather than from eq. (7.131). Useful exact analytical solutions can be obtained only in plane geometry.[18] In fact, one of the drawbacks of the collision probability method is the finite coupling of all the volume elements in the system. Typically the pin cell is divided in less than ten annular volume elements and the method is applicable. In two-dimensional problems, however, the computation of the P_{ijn}'s becomes prohibitively expensive at least in production runs[19] where the main interest is devoted in this book.

Among the techniques devised to shorten the computer running times without causing loss of accuracy, the nodal transmission probability method[20,21] has been demonstrated to be extremely successful. Since the method will later be proposed for two-dimensional lattice cell calculations the present development will also be carried out in more general terms than would be needed for the discussion of the one-dimensional pin cell case.

The transmission probability method allows for anisotropic particle transfer and therefore eq. (7.122) is replaced by

$$\Phi(\mathbf{r}, E, \mathbf{\Omega}) = \int_V K(\mathbf{r}', \mathbf{r}, E, \mathbf{\Omega}) q(\mathbf{r}', E, \mathbf{\Omega})\, d\mathbf{r}' \qquad (7.132)$$

corresponding to eq. (2.67).

The neutron emission density q contains the scattering, fission and extraneous sources.

The method proceeds by discretizing the volume V as is conventional. The coupling of a given volume mesh V_i to the adjacent volume elements V_j is accomplished by means of defining the first flight neutron escape and transmission probabilities. Consider the partial current $J(\mathbf{r}, E, \Delta\mathbf{\Omega})$ into an outgoing direction $\Delta\mathbf{\Omega}$ at surface point $\mathbf{r}$ of the volume element V_i,

$$J(\mathbf{r}, E, \Delta\mathbf{\Omega}) = \int_{\Delta\Omega} J(\mathbf{r}, E, \mathbf{\Omega})\, d\mathbf{\Omega}. \tag{7.133}$$

There are clearly two contributing components in $J(\Delta\mathbf{\Omega})$, one arising from the neutrons born in V_i and escaping at $\mathbf{r}$ without having suffered a collision, and the second part including the neutrons that have traversed V_i after having leaked from one of the adjacent volumes V_j. The balance equation has the form

$$\begin{aligned} J(\mathbf{r}, E, \Delta\mathbf{\Omega}) &= E_i(\mathbf{r}, E, \Delta\mathbf{\Omega}) Q_i(E) V_i \\ &\quad + \sum_j T_j(\mathbf{r}, E, \Delta\mathbf{\Omega}) J_j(E) \end{aligned} \tag{7.134}$$

where $Q_i(E)$ denotes the average source within V_i,

$$Q_i(E) = \int_{V_i} d\mathbf{r} \int d\mathbf{\Omega} q(\mathbf{r}, E, \mathbf{\Omega})/V_i \tag{7.135}$$

and $J_j(E)$ describes the current at the surface S_j into $\Delta\mathbf{\Omega}$,

$$J_j(E) = \int_{S_j} dS \int_{\Delta\Omega} d\mathbf{\Omega} \cdot \mathbf{\Omega} q(\mathbf{r}, E, \mathbf{\Omega}). \tag{7.136}$$

The meaning attached to E_i is the probability for a neutron born in V_i to escape through $\mathbf{r}$ and T_j is the probability for a neutron due to a surface current on S_j to be transferred from V_i without a collision.

Refocusing attention now to cylindrical geometry, eq. (7.134) must be written for the annular cell V_k shown in Fig. 7.7. The partial currents at the surface $r = r_k$ are denoted by $J_k^{\pm}$ depending on whether the inward (−) or outward (+) current is concerned. Using the notation illustrated in Fig. 7.7, eq. (7.134) has the forms[22]

$$J_k^+ = E_k^0 Q_k + T_k^{i0} J_{k-1}^+ + T_k^{00} J_k^- \tag{7.137}$$

and

$$J_{k-1}^- = E_k^i Q_k + T_k^{0i} J_k^- \tag{7.138}$$

where monoenergetic neutrons are assumed. This is, of course, no restriction in a multigroup method. E_k^x denotes the first flight escape probability in the inward ($x = i$) or the outward ($x = 0$) direction. T_k^{xy} are the in–out, out–in or out–out transmission probabilities whose meaning is clarified in Fig. 7.7.

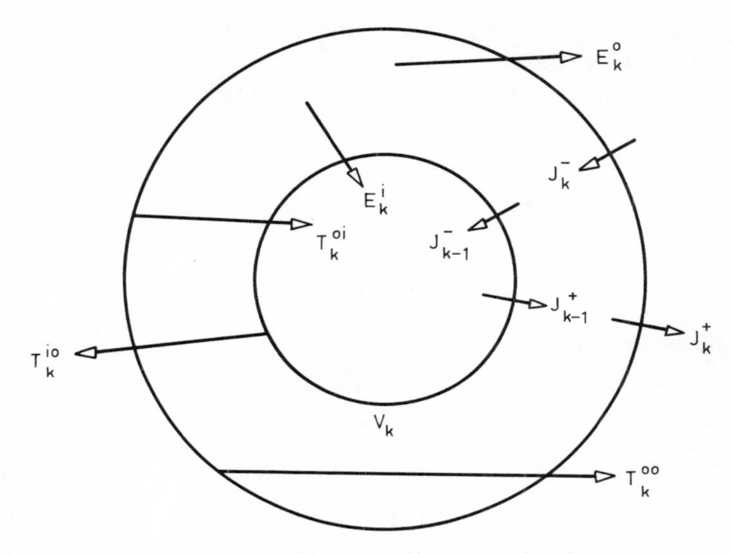

FIG. 7.7. Cell geometry in the transmission probability method.

For the innermost cylinder $k = 1$ only the outward equation

$$J_1^+ = E_1^0 Q_1 + T_1^{00} J_1^- \tag{7.139}$$

possesses physical significance. At the exterior boundary $r = r_I$ the inward group current $J_{I,n}^-$ can be coupled with other energy groups. Neglecting the coupling, the general boundary condition can be formulated as

$$J_I^- = \alpha J_I^+ - L_I \tag{7.140}$$

where α is related to the albedo condition [cf. eq. (2.63)], and L_I is an additional parameter by which the boundary is described.

Equations (7.137)–(7.138) can be solved recursively[23] and in fact eq. (7.138) is nothing else but a recursion relation

$$J_{k-1}^- = a_k^- + b_k^- J_k^-. \tag{7.141}$$

It must be supplemented by another relation of the form

$$J_k^+ = a_k^+ + b_k^+ J_k^-. \tag{7.142}$$

The coefficients a_k^+ and b_k^+ can be solved from the previous equations. By virtue of eq. (7.139) one has

$$a_1^+ = E_1^0 Q_i \tag{7.143}$$

and

$$b_1^+ = T_1^{00}. \tag{7.144}$$

The solution of a_k^+ and b_k^+ yields

$$a_k^+ = E_k^0 Q_k + T_k^{i0}(a_{k-1}^+ + b_{k-1}^+ E_k^i Q_k), \tag{7.145}$$

$$b_k^+ = T_k^{00} + T_k^{i0} b_{k-1}^+ T_k^{0i} \tag{7.146}$$

for all $1 < k \leq I$. Since eqs. (7.138) and (7.141) imply trivially that

$$a_k^- = E_k^i Q_k, \tag{7.147}$$

$$b_k^- = T_k^{0i} \tag{7.148}$$

one has all the recursive coefficients $a_k^\pm$ and $b_k^\pm$ defined in eqs. (7.143)–(7.148). From the boundary condition eq. (7.140) and from eq. (7.142) with $k = I$, one is able to solve J_I,

$$J_I^- = \frac{\alpha a_I^+ - L_I}{1 - \alpha b_I^+} \tag{7.149}$$

and the rest of the partial currents are obtained recursively from eqs. (7.141)–(7.142).

Note that if the energy coupling is included in eq. (7.140), then it should be replaced by a corresponding linear system of equations and the subsequent manipulations should be made in terms of matrices rather than scalars. The development is very similar to the one given above.

The group fluxes within the mesh region V_k are obtained by a local balance equation where the total collision rate $\Sigma_t^k \phi_k V_k$ is set equal to the contribution from the source $Q_k V_k$ and the net current

$$\Sigma_t^k \phi_k V_k = Q_k V_k + (J_k^- - J_k^+) + (J_{k-1}^+ - J_{k-1}^-). \tag{7.150}$$

Since the right-hand side of eq. (7.150) has been computed for each energy group, ϕ_k can be solved directly. The first flight escape and transmission probabilities are calculated *a priori* and tabulated in order to give a convenient basis for interpolation when the problem is run. The main parameters involved in the determination of the E's and T's include the mesh geometry, i.e. the inner and outer radii of the volume V_k together with the total cross-section.

Regardless of whether the collision or the transmission probability methods are employed in the pin cell calculation, the computation is naturally commenced by laying a cylindrical mesh over the unit cell. As indicated in Fig. 7.8, the ordinary Wigner–Seitz equivalence cell of Fig. 2.3 is surrounded by an additional boundary volume where the boundary condition can optionally be implemented.[24] The reflective condition pertaining to a description of an infinite rod array can be used equivalently to filling the boundary volume by an isotropic scattering ring. The latter amounts to the white boundary condition. In view of the collision probability mesh in Fig. 7.6 or the transmission probability mesh in Fig. 7.7, the lattice geometry in Fig. 7.8 can be readily matched with either one of the approximation schemes. In a typical case considered there is a maximum of up to 10–15 discrete mesh volumes.

The pin cell problem is handled as a source problem where the

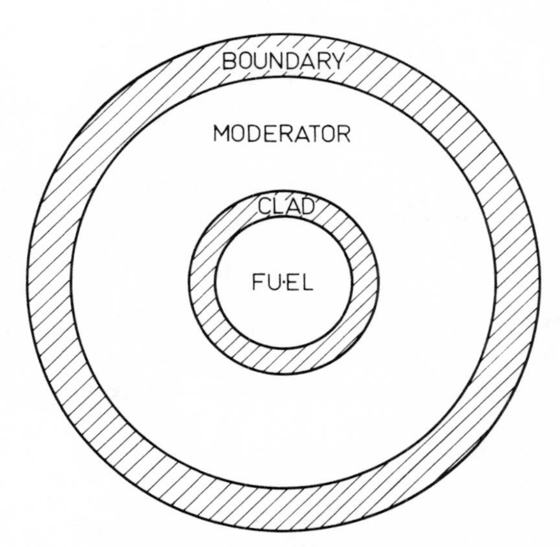

FIG. 7.8. White boundary condition.

relevant scattering and slowing down terms are written explicitly in eqs. (7.123)–(7.124). Also recall that q in eq. (7.132) is comprised of the two parts. As was implied in the flow sheet of Fig. 7.2, the thermal scattering kernel and the multigroup program input of group to group scattering cross-sections are considered in a sub-module of their own and will be discussed later. The slowing down source is calculated in the simplest approximation from the assumption of isotropic elastic scattering. This assumption and the kernel of eq. (7.82) are valid for energies above 1 eV where no chemical binding effects are of importance. The upper energy limit is extended up to about 2 eV in order to include the 1 eV pU^{240} resonance in the thermal spectrum calculation.

Traditionally[15] the energy variable is replaced by speed and therefore eq. (7.82) is cast in the form

$$\Sigma_s(v', v) = \frac{2v\Sigma_s(v')}{(1-\alpha)v'^2} \quad \text{for} \quad \sqrt{\alpha}v' < v < v' \tag{7.151}$$
$$= 0 \quad \text{otherwise.}$$

Correspondingly, the $1/E$ spectrum of eq. (7.80) is expressed via speed and with reference to the thermal cutoff point v_{th}. One has then for the epithermal flux

$$\phi(v) = \phi(v_{th})\frac{v_{th}}{v}. \tag{7.152}$$

The final assumption in calculating the thermalization source term concerns the separability, i.e. in (7.124) the flux is expressed by

$$\phi(\mathbf{r}, v) = Q_{th}(\mathbf{r})\phi(v) \tag{7.153}$$

where $Q_{th}(\mathbf{r})$ describes the spatial variation of the epithermal neutron density.

Letting q_n denote the thermalization source in the nth group (v_{n-1}, v_n)

$$q_n(\mathbf{r}) = \int_{v_{n-1}}^{v_m} Q(\mathbf{r}, v)\, dv \tag{7.154}$$

and substituting eqs. (7.124) and (7.151)–(7.153) into eq. (7.154) one obtains for the group source[15]

$$q_n(\mathbf{r}) = c\,\frac{\Sigma_s(\mathbf{r})Q_{\text{th}}(\mathbf{r})}{(1-\alpha)}\left[\frac{1}{2}(v_n^2 - v_{n-1}^2) - \alpha v_{\text{th}}^2 \ln \frac{v_n}{v_{n-1}}\right] \qquad (7.155)$$

where $\sqrt{\alpha}v_{\text{th}} \leq v_n \leq v_{\text{th}}$ and the constant c is introduced to normalize the total slowing down source. Equation (7.155) can be used as a slowing down source for all elements except hydrogen, where a more careful group source calculation is required. THERMOS or a corresponding code uses up to fifty energy groups, some of which coincide with the MUFT groups of lower energy. A rather accurate method would naturally be obtained by transferring the MUFT slowing down densities to the thermal spectrum code.

As an illustration of THERMOS output, Fig. 7.9 depicts the neutron energy spectrum in a LWR pin cell with mixed U–Pu fuel.[25] Both the 0.3 eV Pu^{239} and the 1.0 eV Pu^{240} resonances are detectable in the figure. The group condensation procedure within

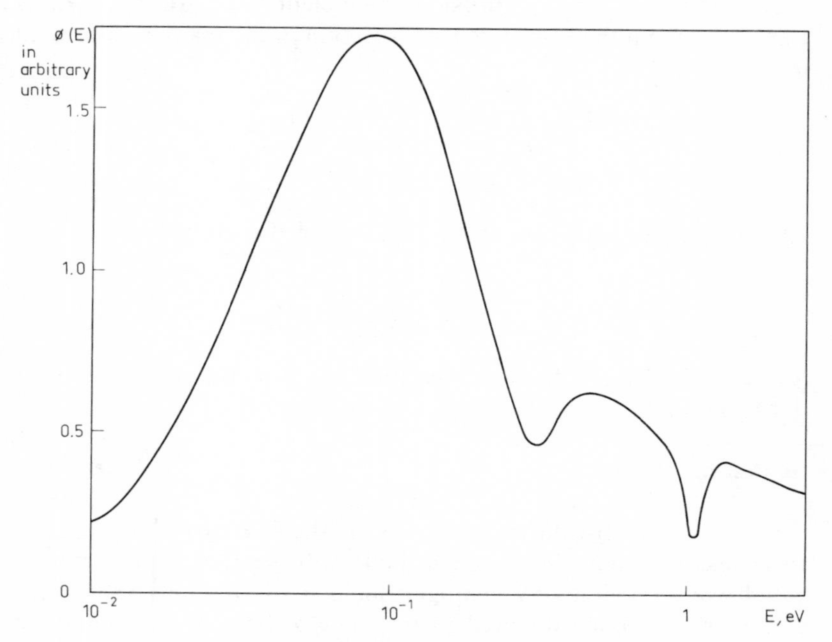

FIG. 7.9. Energy spectrum in a LWR cell with mixed U–Pu fuel.[25]

the thermal energy range is analogous to one applied for the fast group constants [cf. eqs. (7.72)–(7.78)]. One is reminded, however, that the spatial discretization now involves the group constant condensation over the volumes into which the pin cell is divided. The cross-section Σ_x^{cell} is, for example, obtained from

$$\Sigma_x^{\text{cell}} = \frac{\sum_{i=1}^{I} \sum_{n=1}^{N} \Sigma_{xn}^{i} \phi_{in} \Delta v_n V_i}{\sum_{i=1}^{I} \sum_{n=1}^{N} \phi_{in} \Delta v_n V_i}. \tag{7.156}$$

In case thermal few group cross-sections $\Sigma_{x,m}^{\text{cell}}$ are required, the group sum over n in eq. (7.156) is extended only over those multigroups n which are within the broader group interval ΔE_m.

In the isotropic collision probability method described above, a transport correction can be introduced in the total cross-section Σ_t. Replacing Σ_t by $\Sigma_{\text{tr}} = \Sigma_t - \bar{\mu}\Sigma_s$ would be an approximation of the first order. The few group diffusion coefficient D_n^{cell} for the cell is calculated from the transport cross-section in the manner described in section 2.3,

$$D_m^{\text{cell}} = \frac{\sum_n D_n^{\text{cell}} \phi_{\text{cell},n} \Delta v_n}{\sum_n \phi_{\text{cell},n} \Delta v_n} \tag{7.157}$$

where $\phi_{\text{cell},n}$ denotes the group flux within the pin cell and

$$D_n^{\text{cell}} = \frac{1}{3\Sigma_{\text{tr}\,n}^{\text{cell}}} \tag{7.158}$$

[cf. eq. (2.24)]. More accurate methods for the calculation of diffusion coefficients are devised in the literature.[26]

Scattering Laws

One of the drastic differences between the analyses of the fast and thermal energy domains is constituted by the scattering mechanisms which yield qualitatively distinguishable scattering kernels. Extensive work has been devoted to the development of computerized methods which yield the scattering kernel appearing in eq. (7.123).

The interaction of neutrons with the targets can in principle be governed to a far higher degree of accuracy than is required for the present discussion. The theory becomes rather involved[9] and only the main features are sketched below.

The formalism of thermal neutron scattering is based on the scattering law $S(\alpha, \beta, T)$ with the variable α being related to the momentum transfer in a collision

$$\alpha = (E + E' - 2\mu\sqrt{EE'})/AkT \qquad (7.159)$$

and β to the energy transfer

$$\beta = (E - E')/kT. \qquad (7.160)$$

A denotes the ratio of scatterer to neutron mass and μ is the cosine of the scattering angle. $S(\alpha, \beta)$ is tabulated in nuclear data compilations and the scattering kernel $\Sigma_s(E', E, \mu)$ can be derived from S. In fact the microscopic scattering kernel, sometimes referred to as the differential scattering cross-section, is given by[27]

$$\sigma_s(E', E, \mu) = \frac{\sigma_b}{2kT}\sqrt{\frac{E}{E'}}\,e^{-\beta/2}S(\alpha, \beta, T) \qquad (7.161)$$

where σ_b is the scattering cross-section of a bound atom.

The so-called incoherent approximation[9] included in eq. (7.161) implies that the expression is fairly accurate for the inelastic scattering contribution which plays a major part in the energy transfer of the neutron population in a large homogeneous thermal reactor.[9] The elastic cross-sections are edited directly from the data file.[27]

Theoretical models devised for calculation of the scattering law employ usually[3] the intermediate scattering function $\chi(\kappa, t)$ which is a Fourier transform of S. Using conventional variables

$$\epsilon = E - E' \qquad (7.162)$$

and

$$\kappa^2/2 = (E + E' - 2\mu\sqrt{EE'}) \qquad (7.163)$$

the transformation is performed by[28]

$$S(\kappa, \epsilon) = \frac{1}{2\pi}\int_{-\infty}^{\infty} e^{-i\epsilon t}\chi(\kappa, t)\,dt. \qquad (7.164)$$

Various approximations propose the form of χ and the corresponding scattering law is derived from eq. (7.164).

Most concern is directed to a proper consideration of neutron scattering from water. A widely known model due to Nelkin[28] has been used successfully in generating the scattering kernels. The model distinguishes between the translational, rotational and vibrational parameters associated with the dynamics of collision interaction between neutrons and water molecules. The intermediate scattering function is obtained as a product

$$\chi(\kappa, t) = \chi_t(\kappa, t)\chi_r(\kappa, t)\chi_v(\kappa, t). \tag{7.165}$$

The translational, rotational and vibrational terms χ_t, χ_r and χ_v are further given by simple expressions of the form

$$\chi_i(\kappa, t) = \exp\left\{\frac{\kappa^2}{2A}[\gamma(t) - \gamma(0)]\right\} \tag{7.166}$$

where different forms of $\gamma(t)$ apply in the three cases. $\gamma(t)$ is further composed of an integral over a mode frequency function which is allowed to have only discrete delta function terms.[9,28] There are numerous variants and improvements of the Nelkin and related models, e.g. ref. 29. While these result in a more accurate description of scattering from water, there is not much incentive to use them in data production within the domain of the present discussion. For example, the computer running times grow fast when sophistication is increased.

Two-dimensional Cell Calculations

Recalling that the objective in lattice cell calculations is to provide one or two group constants for homogenized fuel assemblies, a decision has to be made concerning the degree of geometric detail that is or has to be included in the analysis. Depending on the heterogeneity within the fuel assembly as well as on the heterogeneity prevailing with respect to the adjacent assemblies, there are different categories of how far the cell geometry is extended.

If the lattice is fairly homogeneous, the smeared group constants calculated for the pin cell can be used for the entire fuel assembly.

Corresponding to this gross method there is the bypass option in Fig. 7.2 which circumvents the two-dimensional calculation at the assembly level.

In many cases the fuel assembly is heterogeneous enough to necessitate a two-dimensional study over the region extracted by symmetry considerations. Most frequently 1/4th or 1/8th symmetries exist in square lattices, whereas 1/3rd, 1/6th or 1/12th is the normal symmetry for hexagonal configuration. The domain boundary corresponds to case A in Fig. 7.6. The heterogeneity may be a result of varying fuel pin enrichment within the assembly or of the flux distortion introduced by control rod insertion and global flux tilts.

The separation of one individual fuel bundle from the array implies physical decoupling of adjacent assemblies. In PWRs where the decoupling cannot always be reasoned to be permissible, the two-dimensional domain may be extended to cover part of the adjacent assembly as shown by case B in Fig. 7.10.

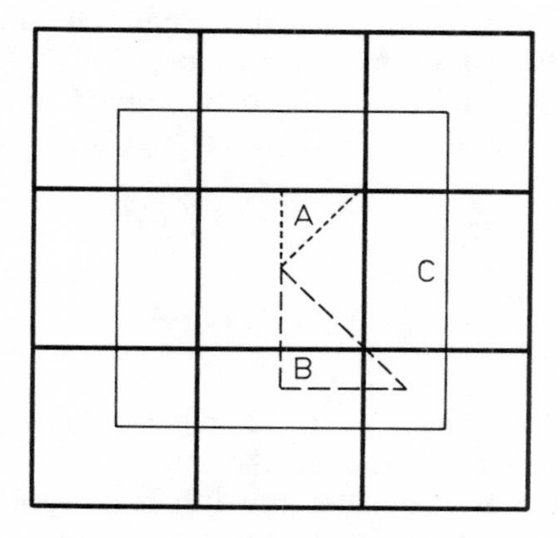

FIG. 7.10. Lattice cells and super-cells.

Frequently the differences in the initial enrichment and burnup of the adjacent fuel assemblies distort the symmetry within the interior fuel assembly. A good example is furnished by locating the nine assemblies of Fig. 7.10 in the core of Fig. 6.6 such that the upper row would consist of assemblies 3,3, and 1, the second row would have assemblies 2,1,1, whereas the lower row would have only two assemblies 3 and 1 in the core and the third position would be extended to the reflector and core barrel. One would now desire to perform a super-cell calculation over region C in Fig. 7.10. Suppose that the 5–10 group condensed cross-section have been calculated for all involved 1/8ths corresponding to A. If these computations were carried out using one spatial mesh point per fuel pin then the super-cell calculation could have one mesh point per up to 8–12 fuel pins and the calculation would not become prohibitively costly. It could possibly be performed for a coarser energy group structure as well.

The two-dimensional calculations are carried out either in the diffusion approximation or using some approximate transport theory method. The extent to which transport theory must be used is limited, because the problem becomes rather large in terms of space-energy mesh points. A PWR assembly would imply one-fourth or one-eighth of a 17×17 spatial grid and an iteration in five to ten energy groups. It is clear that the super-cell considerations are falling beyond the use of transport theory.

The group diffusion equations to be solved over the different optional domains in Fig. 7.10 have in rectangular geometry the form

$$-\left(\frac{\partial}{\partial x} D_n(x, y) \frac{\partial}{\partial x} + \frac{\partial}{\partial y} D_n(x, y) \frac{\partial}{\partial y}\right) \phi_n(x, y) + \Sigma_{tn}(x, y)\phi_n(x, y)$$
$$= \sum_{m=1}^{N} \Sigma_{smn}(x, y)\phi_m(x, y) + \lambda\chi_n \sum_{m=1}^{N} \nu\Sigma_{fm}(x, y)\phi_m(x, y) \tag{7.167}$$

which has been displayed in eq. (2.52) in general geometry. Equation (7.167) is now discretized over the spatial grid and solved by numerical techniques. Consider the ijth mesh point in Fig. 7.11. Equation (7.167) is cast to a finite difference equation by integrating it over the ijth area.[30] In contrast to the procedure of ref. 30, the mesh points are assumed to lie at the centre of a uniform region and not at possible material interfaces. The present discretization method is used in some well-known cell programs, e.g. in ref. 31.

In performing the integration the streaming term is integrated once, converting it into a line integral by using Green's theorem[30]

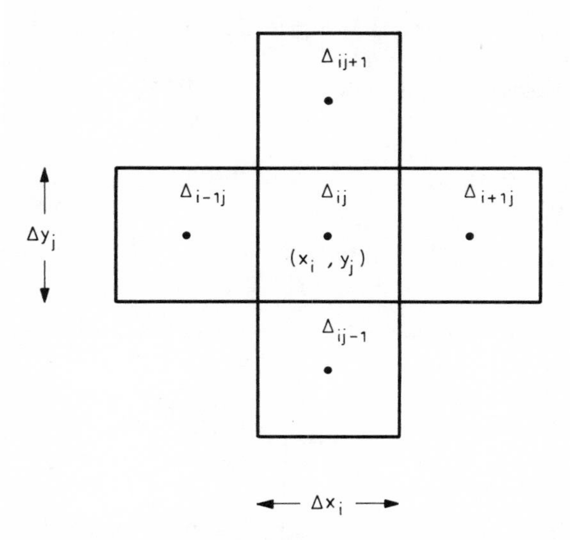

FIG. 7.11. Two-dimensional mesh geometry.

$$\int_{\Delta_{ij}} \left(\frac{\partial}{\partial x} S - \frac{\partial}{\partial y} T\right) dx\, dy = \int_{C_{ij}} (T\, dx + S\, dy) \tag{7.168}$$

where Δ_{ij} is the area of the ijth mesh region and C_{ij} denotes the boundary. The derivative terms become

$$-\int_{\Delta_{ij}} \left[\frac{\partial}{\partial x} D(x, y) \frac{\partial}{\partial x} + \frac{\partial}{\partial y} D(x, y) \frac{\partial}{\partial y}\right] \phi(x, y)\, dx\, dy$$
$$= \int_{C_{ij}} \left[D(x, y) \frac{\partial}{\partial y} \phi(x, y)\, dx - D(x, y) \frac{\partial}{\partial x} \phi(x, y)\, dy\right]. \tag{7.169}$$

In view of eq. (7.169) it is important to observe that although the line integration may be performed along a material interface the integrands $D\mathbf{n} \cdot \nabla\phi$ are unique and well-behaving functions by virtue of the current continuity. Consider, for example, the integral $\int_{\Delta y_j} D(\partial/\partial x)\phi\, dy$ at $x = x_i + (\Delta x_i/2)$. Approximating the integral by

$$\int_{\Delta y_j} D \frac{\partial}{\partial x} \phi\, dy = \left[D(x, y) \frac{\partial}{\partial x} \phi(x, y)\right]_{\substack{y = y_j \\ x = x_i + \Delta x_i/2}} \Delta y_j \tag{7.170}$$

it is the neutron current at the grid line that has to be evaluated. Approaching the interface at $x_i + x_i/2$ along the line $y = y_j$ a linear variation of $\phi(x, y)$ is assumed, i.e.

$$\phi(x, y_j) = \phi(x_i, y_j) + \alpha_i(x - x_i) \qquad (7.171)$$
$$x \in [x_i, x_i + \Delta x_i/2].$$

On the other hand,

$$\phi(x, y_j) = \phi(x_{i+1}, y_j) + \alpha_{i-1}(x - x_{i+1}) \qquad (7.172)$$
$$x \in [x_i + \Delta x_i/2, x_{i+1}].$$

Since the neutron flux is continuous across the mesh boundary one gets

$$\phi_{ij} + \alpha_i \Delta x_i/2 = \phi_{i+1j} - \alpha_{i+1}\Delta x_{i+1}/2 \qquad (7.173)$$

with

$$\phi_{ij} = \phi(x_i, y_j) \qquad (7.174)$$

where the energy group subscripts have been omitted.

In order to determine the current across the interface a fictitious effective diffusion coefficient $D_{i+1/2j}$ is defined by

$$\left[D(x, y)\frac{\partial}{\partial x}\phi(x, y)\right]_{\substack{y=y_j \\ x=x_i+\Delta x_i/2}} = 2D_{i+1/2j}(\phi_{i+1j} - \phi_{ij})/(\Delta x_i + \Delta x_{i+1}). \qquad (7.175)$$

Equation (7.175) implies a central difference scheme[30] for the flux derivative. Deriving the left-hand side of eq. (7.175) from eqs. (7.171)–(7.172) one has

$$D_{ij}\alpha_i = D_{i+1j}\alpha_{i+1} = 2D_{i+1/2j}(\phi_{i+1j} - \phi_{ij})/(\Delta x_i + \Delta x_{i+1}). \qquad (7.176)$$

Solving eqs. (7.173) and (7.176) yields simply[31]

$$D_{i+1/2j} = \frac{\Delta x_i + \Delta x_{i+1}}{\dfrac{\Delta x_i}{D_{ij}} + \dfrac{\Delta x_{i+1}}{D_{i+1j}}}. \qquad (7.177)$$

The integral in eq. (7.170) is evaluated, the result being shown in eqs. (7.175) and (7.177). For the line segments of C_{ij} one proceeds in a manner similar to that indicated above. In the rest of the terms in eq. (7.167) the surface integrals are evaluated from

$$\int_{\Delta_{ij}} f(x, y)\, dx\, dy = f_{ij} \Delta_{ij}. \qquad (7.178)$$

The finite difference diffusion equation is now given by the following five-point formula where the energy group indices are reintroduced

$$-A^n_{ij}\phi_{ni-1j} - B^n_{ij}\phi_{nij-1} + D^n_{ij}\phi_{nij} - F^n_{ij}\phi_{nij+1} - G^n_{ij}\phi_{ni+1j}$$
$$= \left[\sum_{m=1}^{N} \Sigma^{ij}_{smn}\phi_{mij} + \lambda\chi_n \sum_{m=1}^{N} \nu\Sigma^{ij}_{fm}\phi_{mij}\right]\Delta_{ij} \qquad (7.179)$$

with

$$A^n_{ij} = 2\Delta y_j \Big/ \left(\frac{\Delta x_{i-1}}{D_{ni-1j}} + \frac{\Delta x_i}{D_{nij}}\right), \qquad (7.180)$$

$$B^n_{ij} = 2\Delta x_i \Big/ \left(\frac{\Delta y_{j-1}}{D_{nij-1}} + \frac{\Delta y_j}{D_{nij}}\right), \qquad (7.181)$$

$$F^n_{ij} = B^n_{ij-1}, \qquad (7.182)$$

$$G^n_{ij} = A^n_{i+1j}, \qquad (7.183)$$

$$D^n_{ij} = A^n_{ij} + B^n_{ij} + F^n_{ij} + G^n_{ij} + \Sigma^{ij}_{tn}\phi_{nij}\Delta_{ij}. \qquad (7.184)$$

In eqs. (7.180)–(7.183) D_{nij} denotes the value of the diffusion coefficient in the nth group at (x_i, y_j).

At the boundaries eq. (7.184) is altered, accounting for the boundary condition imposed. The discretization of the general boundary condition in eq. (2.58) follows the procedure given above. In the line integrals over the exterior or symmetry interfaces the appropriate boundary condition can be inserted in a trivial manner.[30]

One of the characteristics in the derivation was embodied in the harmonic weighting of the diffusion coefficient in eq. (7.177). Letting all the mesh spacings be equal, one observes that $D_{i+1/2j} = 2D_{ij}D_{i+1j}/(D_{ij} + D_{i+1j})$, whereas simple averaging would yield just $1/2(D_{ij} + D_{i+1j})$.

The five-point difference representation is essentially tantamount to the first order flux approximation in eqs. (7.171)–(7.172). A Taylor expansion of a higher order would incorporate coupling to a larger number of mesh points. The purely numerical part of the solution will be deferred to section 7.6, where iteration techniques are discussed.

Diffusion theory methods are expected to fail in the treatment of heavily absorbing regions. The analysis would there require certain precalculated boundary conditions essentially of the type of albedo matrices or accurate extrapolation distances. A somewhat different approach is used in the British WIMS-scheme[(31)] where the black area, i.e. heavy absorber, is represented by an effective diffusion coefficient. From the results of a super-cell transport calculation one obtains the diffusion coefficient at the black surface. Knowing the diffusion coefficient for the surrounding moderating medium the effective diffusion coefficient of the black region is obtained essentially from eq. (7.177) where $D_{i+1/2}$ would represent the interface value and D_i the value in the surrounding medium. Specifying the thickness of the black volume x_{i+1}, one obtains the unknown quantity which would be now denoted by D_{i+1}.

A more precise statement of the recipe can be made by encircling the black area by a line with perimeter L. The total net current J_s corresponding to the transport solution is obtained from

$$\int_L D_S \mathbf{n} \cdot \nabla\phi(\mathbf{r})\, dl = J_s \tag{7.185}$$

where $\mathbf{n}$ denotes the outward normal of the black surface. Consider an absorber pin within a regular pin array with spacing p. Letting the flux gradient be constant and equal to $\Delta\phi/p$ the integral in eq. (7.185) can be evaluated trivially to give[(31)]

$$D_S = \frac{J_s}{4\Delta\phi}. \tag{7.186}$$

If D_F denotes the known diffusion coefficient of the surrounding fuelled pin cells and D_p that of the poison pin cell then one gets from eq. (7.177)

$$D_p = \left(\frac{2}{D_S} - \frac{1}{D_F}\right)^{-1} \tag{7.187}$$

where D_S is inserted from eq. (7.186). Note that the problem studied above bears relevance in practical reactor lattices in conjunction with either burnable poison fuel pins or PWR rod cluster control assemblies.

In BWRs where the control devices are cruciform absorber blades the detailed geometry is first homogenized so that the blade is represented in one-dimensional slab geometry. Paraphrasing again the WIMS technique[31] D_S is now calculated directly from the definition

$$J_S = 2D_S \frac{\phi_F - \phi_P}{(\Delta x_F + \Delta x_p)} \tag{7.188}$$

where x_p and x_F denote the thicknesses of the homogenized poison slab and the adjacent medium, respectively. For D_p one obtains from eq. (7.177)

$$D_p = \Delta x_p \left(\frac{\Delta x_F + \Delta x_p}{D_S} - \frac{\Delta x_F}{D_F} \right)^{-1}. \tag{7.189}$$

Recall that J_S, ϕ_F and ϕ_p are assumed to be known on the basis of a one-dimensional transport calculation performed *a priori.*

While transport theory is exclusively used in the thermal domain of the pin cell study, there have not been many rapid two-dimensional algorithms in existence to make transport calculations feasible on the different optional assembly regions depicted in Fig. 7.10. Among the methods that conceivably would be amenable to production runs in core fuel management there has traditionally existed a competition between the S_N method[9] based on the integro-differential form and the collision probability method which is a straightforward extension of the corresponding one-dimensional technique covered earlier in this section. Regarding both the computer running times and accuracy there are clear indications[21] that the transmission probability method would be more efficient than either the S_N or the collision probability method in two-dimensional fuel assembly calculations. The technique has been discussed previously in one dimension and the two-dimensional analogy would proceed directly from eq. (7.134).[20,21]

Concerning the judgement of whether one should resort to transport theory versions, it can be hardly insisted that the overall analytic assessment capability of a power utility could or should be comprehensive enough to guarantee the same remarkable accuracy as maintained in the transport solution for a fuel assembly. A

1·087 1·118							
0·937 0·954	1·050 1·059						
1·157 1·174	1·130 1·139	P 0·404 0·437	Diagonal symmetry axis				
1·084 1·099	1·164 1·165	0·926 0·916	0·888 0·866				
1·070 1·084	1·169 1·175	0·976 0·960	0·890 0·866	0·868 0·845			
1·096 1·112	1·123 1·127	0·991 0·977	0·852 0·836	0·871 0·848	0·925 0·907		
1·167 1·187	0·936 0·934	1·040 1·035	P 0·387 0·414	0·907 0·894	1·017 1·004	1·046 1·045	
0·933 0·944	1·069 1·074	1·168 1·162	1·102 1·093	1·121 1·107	1·112 1·106	0·941 0·938	1·067 1·069

FIG. 7.12. Power distributions for a BWR assembly with burnable poison rods. Reproduced from ref. 21 with permission of the American Nuclear Society.

counterargument may be established by considering certain numerical benchmark problems where diffusion theory is expected to be inadequate. In Fig. 7.12 a power distribution is shown over a BWR fuel assembly where there are three burnable poison rods. By virtue of symmetry only one-half of the assembly needs to be drawn. The results of Fig. 7.12 were originally presented in ref. 21. The poison rods can be easily identified since they cause a flux depression where the power level is less than one-half of that in any other pin cell and below 40% of the average. Comparison of the transmission probability and diffusion theory methods shows that there is a severe discrepancy at the poison locations. However, some of the error will be smoothed out when the homogenized group constants are derived by spatial and energy averaging procedures.

Energy Point Methods

In lattice cell calculations and particularly in the thermal energy domain conventional multigroup schemes lead to rather expensive computer runs which have to be repeated quite often. While some of the developments in the spatial discretization have been discussed earlier, only the multigroup scheme has been proposed for the description of the energy, lethargy or speed variables. A method based on discrete energy points rather than energy intervals has been devised[32] for rapid computation of thermal spectra and group constants.

In the energy point method the group fluxes ϕ_n and the group parameters Σ_{xn} appearing in the rest of this book are replaced by corresponding values now assigned to the discrete energy point E_n. The flux weighted integrals are handled by the principle of weighting the energy points, i.e. ϕ_n now becomes defined by

$$\phi_n = w_n \phi(E_n) \tag{7.190}$$

where w_n represents the weight function and in a sense is analogous to the group width ΔE_n.

In calculating reaction rates of the type

$$\begin{aligned} R_x &= \int \Sigma_x(E)\phi(E)\,dE \\ &= \sum_n w_n \Sigma_x(E_n)\phi(E_n) \end{aligned} \tag{7.191}$$

in the vicinity of resonances it is customary[32] to change the variable of integration such that coarse mesh integration formulae can be still employed.

To illustrate the improvements that can be achieved by the use of more advanced methods, some results of LWR lattice cell calculations are shown in Table 7.5.[22, 33] The THERMOS[15] scheme employs the collision probability method and 48 energy groups, whereas in FACEL calculations the transmission probability and 24 energy points are used.

The thermal cutoff energy is placed at 2.5 eV in the results tabulated in Table 7.5. The fuel depletion is not regarded in the pin cell calculation, but the isotopic composition is given as input from the modules that will be described in the following section.

The comparison indicates that the FACEL values are generally very close to the basically more accurate THERMOS results. At no burnup stage is the discrepancy large enough to make the more

TABLE 7.5
Relative Thermal Absorption Rates in a Typical LWR Pin Cell[33]

Isotope	Burn-up GWd/tU	FACEL	THERMOS
	0	0.8461	0.8448
	10	0.4903	0.4893
U^{235}	20	0.3282	0.3280
	30	0.2195	0.2195
	40	0.1309	0.1311
	0	0.1032	0.1035
	10	0.0914	0.0919
U^{238}	20	0.0915	0.0921
	30	0.0952	0.0958
	40	0.1012	0.1012
	10	0.2911	0.2901
	20	0.3770	0.3760
Pu^{239}	30	0.4194	0.4189
	40	0.4464	0.4467
	10	0.0611	0.0619
	20	0.1018	0.1016
Pu^{240}	30	0.1296	0.1286
	40	0.1518	0.1502
	10	0.0205	0.0205
	20	0.0557	0.0557
Pu^{241}	30	0.1296	0.1286
	40	0.1183	0.1186

costly THERMOS runs necessary in the fuel management work. A highly relevant aspect of comparison is concerned with the computer running time. In the computing environment of ref. 33 the THERMOS execution times are longer by a factor of the order of 20.

7.4. Depletion Routines

In the planning systems where fuel depletion is described implicitly and the fuel assembly cross-sections are calculated and tabulated as functions of burnup, the block 3 in Fig. 7.1 actually belongs to the lattice cell module. Given the initial isotopic number densities N^m of the materials present in the lattice cell, the objective of the burnup module is to compute the depletion effects as a function of time.

The main input to the depletion calculations consists of the total flux $\phi(\mathbf{r})$ calculated in section 7.3 and of the homogenized group cross-sections. The flux level is specified in principle for the isotopic composition that prevails at the beginning of the burnup step. This will be implied by the notation $\phi(\mathbf{r}, t)$. Frequently only a few special fuel pins are considered at pin level and in case the homogenization has been performed by two-dimensional methods the smeared number densities are given regionwise pursuant to the $x - y$ mesh used.[31] The change of the number density N^m is governed for each $\mathbf{r}$ and t by the equation

$$\frac{dN^m}{dt} = -(A_m + \lambda_m)N^m + \sum_k \alpha_{km} C_k N^k + \sum_k \beta_{km} \lambda_k N^k + \sum_k y_{km} F_k N^k \tag{7.192}$$

where

$$A_m = \sum_n \sigma_{an}^m \phi_n, \tag{7.193}$$

$$C_m = \sum_n \sigma_{\gamma n}^m \phi_n, \tag{7.194}$$

$$F_m = \sum_n \sigma_{fn}^m \phi_n \tag{7.195}$$

are sums over all the group absorption capture and fission rates, respectively. λ_k denotes the decay constant of the element k and the parameters α_{km} and β_{km} are equal to zero or unity depending on whether the species m is a capture or decay product of the component k. The last term of eq. (7.192) describes the fission yield and applies therefore only to the fission products m. The space dependent group fluxes appearing in eqs. (7.193)–(7.195) are in careful analyses[31] made to include the effect of the control mode applied.

Expressing eq. (7.192) by

$$\frac{dN^m}{dt} = \sum_k \gamma_{km} N^k \tag{7.196}$$

and observing that $N^m(\mathbf{r}, t)$ is known at the beginning of the time step, these entries are burnt up over the interval Δt by

$$N(\mathbf{r}, t+\Delta t) = N(\mathbf{r}, t) + \Delta t \sum_k \gamma_{km}(\mathbf{r}, t) N^k(\mathbf{r}, t). \qquad (7.197)$$

Equation (7.197) is merely a statement of the integration process that goes into it. The integration can be carried out either numerically or analytically and an outline of both techniques will be sketched below. There is no clear preference as to which of the all plausible methods would be most efficient. A short review on the topic is included in ref. 34. Evidently there is no unique way to decide the optimal technique which is dependent on the extent of the fission product library maintained. Typically, m would vary over 15–25 heavy or combustible nuclides and over 25–50 fission products.

In order to display the concept of numerical integration employed in the popular LASER code,[24] rewrite eq. (7.192) as

$$\frac{dN^m}{dt} = -D_m N^m + \sum_k \gamma^{km} N^k \qquad (7.198)$$

where all the source contribution is lumped in the coefficients γ^{km}, $k \neq m$. The destriction factor D_m is simply

$$D_m = A_m + \lambda_m. \qquad (7.199)$$

Integrating eq. (7.198) yields

$$N^m(t) = R_m^-(t) \int_0^t \sum_k \gamma^{km}(t') N^k(t') R_m^+(t')\, dt + N_0^m R_m^-(t) \qquad (7.200)$$

where $R_m^{\pm}$ denote the exponential integrals

$$R_m^{\pm}(t) = \exp\left[\pm \int_0^t D_m(t')\, dt'\right] \qquad (7.201)$$

and

$$N_0^m = N^m(0). \qquad (7.202)$$

Clearly, the time scale can always be shifted to make eq. (7.200) valid for an arbitrary time step. The method used in LASER considers eq. (7.200) for a fixed k and the entire decay chain for the pair of indices k and m. Commencing from the first nuclide of the chain for which no precursors exist, eq. (7.200) is applied from one nuclide to another until the species m. The integrand in eq. (7.200) is replaced by a recursive expression which is computed numerically.

Among the variety of analytic solution techniques the method of linear chains is used in certain computer codes.[34, 35] There the chains are constructed somewhat fictitiously so that the quantity N^m will in general represent just a fraction of a real physical number density. At each branch point of a decay chain the linear method separates one path per each chain and the general expression for the nth stage becomes

$$\frac{dN^n}{dt} = Y_n + S_{n-1}N^{n-1} - D_nN^n \tag{7.203}$$

where S_n denotes the source from capture or radioactive decay and Y_n is the fission yield. In other words, S_n comprises the second and third terms in eq. (7.192), while Y_n designates the last term. The sums in eq. (7.192) have to be taken pursuant to the linear chain approach. For the chain precursor $n = 1$ in eq. (7.203) and the S_0 term does not exist. Equation (7.203) describes a linear system where n varies over the entire linear chain. The equations are solved conveniently by Laplace transform techniques and the result can be written in the form[35]

$$N^n(t) = \sum_{m=1}^{n} \frac{1}{S_n} \prod_{k=m}^{n} S_k \left\{ Y_m \left[\frac{1}{\prod_{l=m}^{n} D_e} - \sum_{j=m}^{i} \frac{\exp(-D_j t)}{D_j \prod_{\substack{i=m \\ i \neq j}}^{n} (D_i - D_j)} \right] + N^m(0) \left[\sum_{j=m}^{n} \frac{\exp(-D_j t)}{\prod_{\substack{i=m \\ i \neq j}}^{n} (D_i - D_j)} \right] \right\}. \tag{7.204}$$

The results from a lattice cell calculation have to be updated after a burnup interval of about 2–5 GWd/tU. Computer algorithms are provided usually with facilities to accomplish flexible transfer between the lattice cell and fuel depletion modules. In view of Fig. 1.15 and Table 7.4 it is seen that heavy isotopes have largely different absorption quotas over the typical LWR burnup range. Absorption in U^{238} remains almost constant, whereas U^{235} and Pu isotopes may exhibit large variations of relative importance even over an interval of 5 GWd/tU.

At the depletion stage there will arise situations where the cell averaged cross-sections of fissionable isotopes and burnable poisons have to be modified in order to account for self-shielding effects. These are a consequence of heavy absorption on the fuel surface which results in a flux depression in the interior fuel pin. Pronounced resonance absorption is an example where the so-called self-shielding factors are introduced in the effective absorption cross-sections.[9, 31]

7.5. Reactor Simulation

Core performance analysis constitutes an essential phase in core management activities. The dotted link of the flow diagram in Fig. 7.1 is relevant only in the high degree calculations with explicit burnup consideration and may be totally ignored in the practical fuel management work discussed in this book. Consequently assembly-averaged group constants as functions of fuel depletion, coolant void fraction and neutron spectrum averaged burnup are needed as input to the reactor simulator module. The output consists mainly of the neutron flux and core power distributions pertaining to a desired stage of the core history, while a flexible editing of various other parameters is needed for deriving the safety related thermal margins and feedback effects. Based on the results of the digital core simulation, feasible options of loading patterns and control strategies can be specified within the broader scope discussed in Chapter 6. In other words, the output is employed in the logical decisions where the fuel management strategies are determined over a few core cycles.

In LWRs with short neutron mean free paths and with a large core in terms of neutron diffusion lengths the spatial analysis must use a rather coarse discretization mesh. Radially one would not allow more than one mesh point per a BWR fuel assembly[36] and one to four points per a PWR fuel assembly,[37] whereas the axial mesh spacing is even larger. One is obliged to accommodate distances of the order of 20 cm between the mesh points. No more than two energy groups are used or even needed in LWRs. In fact, the most popular schemes employ only one group. The inaccuracies will then arise principally at the core–reflector interface (cf. the discussion in conjunction with Fig. 2.5), and to some extent at the interior core interfaces between the fuel assemblies which possess conspicuously different irradiation histories.

The major tasks performed by a digital reactor simulation code are indicated in Fig. 7.13. The calculation of the neutron fluxes and the criticality eigenvalue occupies most interest and the subsequent discussion of this section will be conducted around this problem. In spite of some salient differences between BWRs and PWRs the computational techniques can be demonstrated within the same framework, recalling, however, that the routes in Fig. 7.13 may be

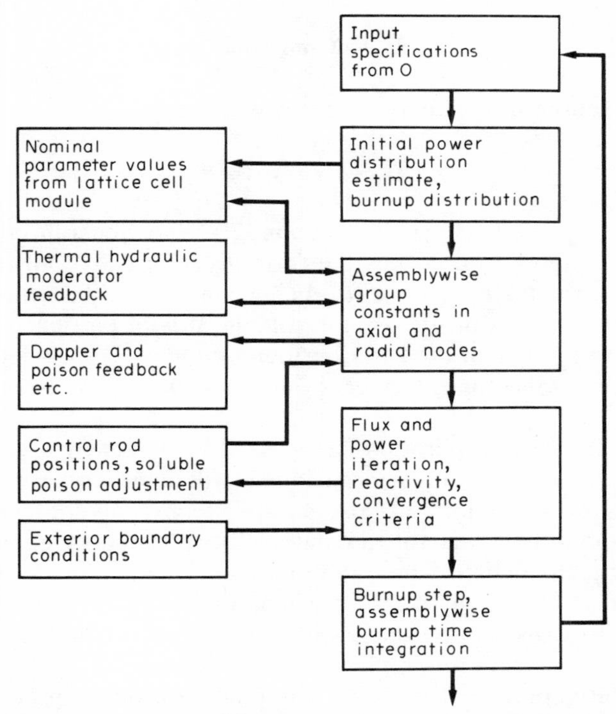

FIG. 7.13. Flow diagram of global reactor calculations (cf. block 4 in Fig. 7.1).

traversed with different frequencies in the two cases. The axial iteration between the flux and moderator voidage requires far more concern in BWRs, whereas the axial uniformness of the PWR flux distribution relaxes partially the need to perform three-dimensional calculations for other than survey purposes.

Finite Difference Methods

The two-group form of the diffusion equation is written as

$$-\nabla \cdot D_n(\mathbf{r})\nabla\phi_n(\mathbf{r}) + \Sigma_{rn}(\mathbf{r})\phi_n(\mathbf{r}) = q_n(\mathbf{r}) \tag{7.205}$$

where [cf. eq. (2.52)] the fast group effective source q_1 is defined by

$$q_1 = \lambda\chi_1(\nu\Sigma_{f1}\phi_1 + \nu\Sigma_{f2}\phi_2) \tag{7.206}$$

and the source in the thermal group is given by

$$q_2 = \lambda\chi_2(\nu\Sigma_{f1}\phi_1 + \nu\Sigma_{f2}\phi_2) + \Sigma_{s12}\phi_1. \tag{7.207}$$

Pursuant to the analysis in section 2.5, the general boundary condition imposed on ϕ_n has the form given in eq. (2.58) with the specific variants for a_n and b_n being applied at the exterior or symmetry surfaces of the reactor volume. It is of particular importance to reiterate that the exterior boundary may be specified either at the core–reflector interface or at the outer boundary of the reflector.

Since the power iteration between the two energy groups is encountered on many other occasions it will be described generally in the following section. It is also possible to circumvent the regular power iteration in reactor simulation by introducing appropriate transient techniques[36] where the thermal group amplitude more or less follows the form of the fast amplitude. It is therefore most relevant to consider the one-group form of eq. (7.205) where the removal cross-section is replaced by absorption Σ_a.

The derivation of the finite difference formulae proceeds in analogy with the treatment of the two-dimensional case in section

7.3. The integration is now naturally to be extended over the appropriate three-dimensional nodes. To avoid the intricacy of multiple indices, drop all the group notation in the equations and designate by 0 the node of interest. Correspondingly, let ϕ_0 denote the flux value assigned to the centre of the node and let ϕ_i $(i = 1, 2, \ldots, 6)$ denote the fluxes at the six adjacent nodes, four horizontal and two vertical neighbours designated by 1, 2, 3. Letting further the horizontal grid be uniform with spacing h and similarly l denoting the vertical spacing the three-dimensional finite difference equation obtains the form

$$C_{00}\phi_0 - \sum_{i=1}^{6} C_{i0}\phi_i + \Sigma_{ao}\phi_0 = q_0 \tag{7.208}$$

where

$$C_{i0} = \frac{2}{h^2}\frac{D_i D_0}{D_i + D_0} \tag{7.209}$$

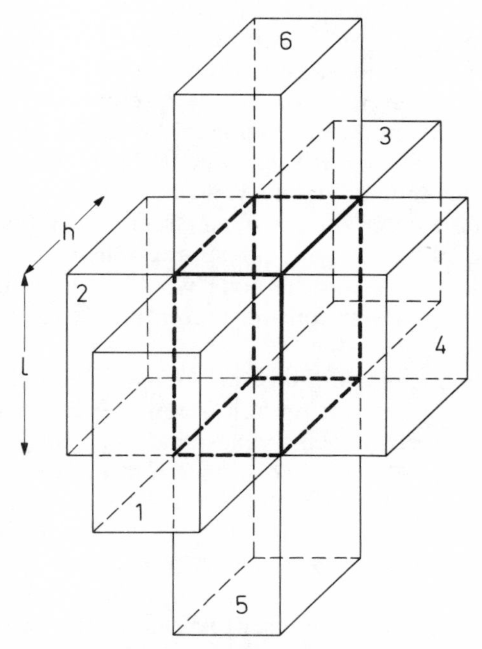

FIG. 7.14. Three-dimensional mesh geometry.

if i corresponds to one of the four horizontally adjacent nodes, $i = 1, 2, 3, 4$ using the notation in Fig. 7.14, and

$$C_{i0} = \frac{2}{l^2} \frac{D_i D_0}{D_i + D_0} \tag{7.210}$$

for the two vertically contiguous nodes. Equivalently to eq. (7.184) C_{00} is obtained from

$$C_{00} = \sum_{i=1}^{6} C_{i0}. \tag{7.211}$$

Equation (7.208) is an immediate extension of eq. (7.179) into three dimensions observing, however, that uniform grids are assumed here both horizontally and vertically. In order to derive eq. (7.208) one would proceed as for eq. (7.179) and then divide the resulting equation by the node volume $h^2 l$.

The numerical solution of the difference equation (7.208) employs standard techniques which can be shown to converge.[30] The discussion of the techniques is deferred to section 7.6, while it is of interest to comment on the physical interpretation of the flux values ϕ_0. As is obvious from the derivation in the earlier two-dimensional case, ϕ_0 denotes the flux at the centre point of the node 0. Some other existing techniques would yield the average value over the entire node. In the latter approach, an effective reaction rate for the node could be obtained with no appreciable error at all. Then, however, the leakage from the node would include an error large enough to make the point concept preferable, at least according to certain studies performed.[36, 37]

Once the numerical solution of eq. (7.208) is completed over the entire core and the flux distribution is converged, a test is made on whether the core eigenvalue corresponds critically. If not, then either the control rods are repositioned and updated homogenized cross-section sets are calculated for the axial nodes where rod positions have an effect or the soluble poison concentration is altered. The particular loop to accomplish this is included in Fig. 7.13. Since the soluble boron control is uniform, it is usually incorporated in the formalism itself. Excluding the boron absorption Σ_{aB} from Σ_a, eq. (7.208) is rewritten as

$$C_{00}\phi_0 - \sum_{i=1}^{6} C_{i0}\phi_{i0} + \Sigma_{a0}\phi_0 - \frac{1}{k}\,\nu\Sigma_{f0}\phi_0 = -\,c_B\,\Sigma_{aB}\phi_0 \qquad (7.212)$$

where c_B denotes the boron concentration. Recalling that in the right-hand side of eq. (7.208)

$$q_0 = \frac{1}{k}\,\nu\Sigma_{f0}\phi_0 \qquad (7.213)$$

it is seen that the critical boron concentration c_B can be solved from eq. (7.212) in an equivalent manner as the multiplication factor k is iterated from eq. (7.208).

Since both eqs. (7.208) and (7.212) are effectively homogeneous equations the neutron flux can be normalized arbitrarily. However, the reactor is operated at some desired power P and consequently the normalization is done according to eq. (1.28). This implies that the core integrated power generation density is set equal to P.

The core simulation incorporates a series of determinations of criticality and power distribution over the fuel cycle at desired burnup intervals. For any given burnup step the flux is either assumed to be constant or allowed to vary in some predicted way and the burnup integral is computed from a discretized form of eq. (1.27). This is done for each node and new group constants pertaining to the succeeding step are then looked for from the parameter tabulations precomputed by a lattice cell computer code.

Methods of Higher Order

While the direct finite difference methods are still in common use, there prevail certain incentives to apply more accurate methods. Firstly, the crude first order difference schemes coupled with the use of coarse meshes may become intolerably inaccurate. They cannot be regarded totally self-sufficient, but subsidiary calculations with more refined grids are necessary for systems where noticeable heterogeneities are present. Mesh refinement becomes prohibitive rather immediately, since the number of unknown flux values grows rapidly with decreasing the mesh spacing in a three-dimensional grid. More advanced techniques are discussed, for example, in a review article by Henry.[38] Here a more detailed discussion will be devoted only to the finite element method which has shown promise for application in core analysis.

The finite element method belongs to the family of variational approximation schemes with particular polynomial trial functions.[39] Ignoring the group index in eq. (7.205) the one-group diffusion equation is written formally as

$$(B\phi)(\mathbf{r}) = q(\mathbf{r}) \tag{7.214}$$

with

$$(B\phi)(\mathbf{r}) = -\nabla \cdot D(\mathbf{r})\nabla\phi(\mathbf{r}) + \Sigma_a(\mathbf{r})\phi(\mathbf{r}). \tag{7.215}$$

Associated with eq. (7.214) a quadratic functional

$$F(\phi) = (B\phi, \phi) - 2(\phi, q) \tag{7.216}$$

is introduced in the variational methods. The variational principle states[39] that minimizing the functional $F(\phi)$ in eq. (7.216) is equivalent to seeking the solution of eq. (7.214).

Pursuant to eq. (4.25) the inner product appearing on the right-hand side of eq. (7.216) is defined as integration over the reactor volume V. Written out explicitly

$$F(\phi) = \int_V (D\nabla\phi \cdot \nabla\phi + \Sigma_a f^2 - 2\phi q)\, d\mathbf{r} - \int_S D\phi\mathbf{n} \cdot \nabla\phi\, dS. \tag{7.217}$$

Partial integration has been performed in the streaming term in eq. (7.217) and the use of Gauss's theorem generates the surface integral term. It is observed that the boundary conditions at S can be inserted directly in the surface term. In particular, the surface integral vanishes at a free boundary where $\phi = 0$.

The reactor volume is partitioned into spatial nodes referred to now as finite elements. The discretization can follow the sub-assembly boundaries but it is by no means limited to that and, in fact, in rectangular lattice configuration one could easily employ triangularization. It has the utility of facilitating local mesh refinement more readily than an orthogonal grid would do. Within each subdivision k a set of trial functions $\{u_i^k(\mathbf{r})\}$ is defined such that u_i^k is a polynomial in the spatial coordinates vanishing outside the element. The flux is approximated in the form

$$\phi(\mathbf{r}) = \sum_k \sum_i a_i^k u_i^k(\mathbf{r}) \tag{7.218}$$

which expression is substituted in the variational functional in eq. (7.127). The stationary point of F where the partial derivatives $\partial F/\partial a_i^k$ vanish for all k and i renders the finite element expansion coefficients a_i^k and ultimately the approximate solution ϕ from eq. (7.218). For the coefficients a_i^k a linear system of equations is obtained in the form

$$\sum_k \sum_i (Bu_i^k, u_j^l)a_i^k = (q, u_j^l) \tag{7.219}$$

for all l and j. By virtue of the localized trial functions the transformation matrix is sparse, i.e. a great deal of the inner products on the left-hand side of eq. (7.219) are equal to zero.

The finite element method possesses theoretically acceptable convergence properties. Numerical experimentation in some two-dimensional cases on realistic power reactor cores has confirmed the potential of the method and established its superiority to conventional finite difference schemes (cf. ref. 40, for example). Some concern has to be devoted to the singular points at the intersections of mesh lines where two material interfaces cross each other.

The element functions $u_i^k(\mathbf{r})$ are constructed of polynomials and defining them properly as unit functions the expansion coefficient a_i^k will have the physical meaning of representing the values of the flux and its derivatives. Depending on whether the meaning of pointwise flux values within the node or the flux and current values at the node boundaries is assigned, the element functions can be chosen to be products of one-dimensional Lagrange or Hermite polynomials, respectively. The inherent flux and current continuities can be directly imposed in the Hermitian scheme, which therefore is expected to have a more promising potential in coarse mesh applications.

A linear finite element scheme where u_i^k's are linear functions is rather closely related to the regular finite difference approach.[39] The more accurate element approximations entail linear transformations of higher rank. Robinson and Eckard[41] have proposed a one-group finite difference technique which in spite of higher degree reduces to the similar contiguous coupling as the one involved in eq. (7.208). Their method has been shown to agree sufficiently well with more accurate results as well as with experiment. Accuracies of the order of 6% have been demonstrated by this method. It is of interest to note that the coarse mesh running time is 2 min on a modern computer as compared with several hours in the case of a fine mesh conventional difference method.[41]

As an example of the power maps edited from a reactor simulator a normalized power distribution over one-eighth of a PWR core[42] is shown in Fig. 7.15. The calculation is performed near the middle of a cycle and at full power. The two-dimensional cut is representative in this case where a large PWR with no inserted control rods is involved.

A number of three-dimensional approximation schemes are based on synthesis of the axial distribution with precomputed two-dimensional radial flux maps. Synthesis methods have shown some promise and they might have deserved a treatment here. However,

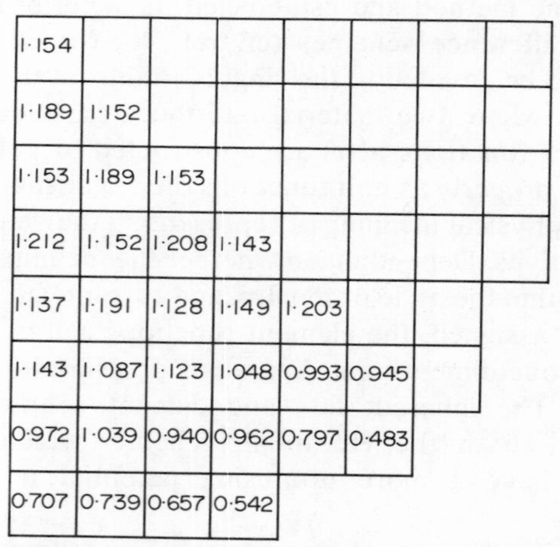

1·154							
1·189	1·152						
1·153	1·189	1·153					
1·212	1·152	1·208	1·143				
1·137	1·191	1·128	1·149	1·203			
1·143	1·087	1·123	1·048	0·993	0·945		
0·972	1·039	0·940	0·962	0·797	0·483		
0·707	0·739	0·657	0·542				

FIG. 7.15. Normalized assemblywise power distribution in a PWR core.[42] One-eighth symmetry.

the basic flux modelling and method of solution are founded on very much the same routines as in the finite difference or finite element methods.[38, 43] The codes seem to be tailormade to an extent which is seen to fall outside the scope of the present discussion.

Nodal Methods

Among the chronologically first approaches devised for digital reactor simulation a nodal technique frequently referred to as the FLARE method[44] has been a popular starting point for a variety of computer codes. Subsequent improvements[45] have made FLARE schemes to widely adopted tools and a terse description of the method is therefore appropriate.

The FLARE equations can be deduced from the discretized finite difference equations.[38] However, the method does not involve the solution of the differential equation (7.208), but the original equations are constructed from the node-to-node neutron transfer relations. Letting Q_i denote the rate at which neutrons are injected in the mode i and letting W_{ij} denote the probability that a neutron born within the node i will be absorbed at j, one obtains the following balance equation:

$$R_{aj} = \sum_i W_{ij} Q_i \tag{7.220}$$

for calculating the absorption rate R_{aj} at the node j. Usually the sum in eq. (7.220) is extended only over the six contiguous neighbours of the node. To establish an equation relating R_i and Q_i one is reminded of the fact that the nodes are very large in comparison with the diffusion lengths. Therefore the simple interpretation of the effective multiplication factor k_∞ for an infinite lattice can be used in writing the first order relation

$$Q_i = k_{\infty i} R_{ai}. \tag{7.221}$$

Combining eqs. (7.220) and (7.221) one obtains the basic FLARE equation for the node 0,

$$Q_0 = \frac{k_{\infty 0} \sum_{i=1}^{6} W_{i0} Q_i}{1 - k_{\infty 0} W_{00}} \tag{7.222}$$

where the node labelling of Fig. 7.14 has been used. Once the values of $k_{\infty i}$ and W_{ij} are known throughout the core, the reaction rate solution can be iterated from eq. (7.222).

Concerning the derivation of the W_{ij}'s, numerical methods of collision theory could be applied in principle, whereas convenient analytical methods have been devised as well.[45] An elementary derivation is based on the simplified critical one-group diffusion equation expressed by

$$-\nabla^2\phi + \frac{k_\infty - 1}{M^2}\phi = 0 \tag{7.223}$$

where in comparison with eq. (7.205) the diffusion coefficient is assumed to be space-independent and M^2 denotes the migration area

$$M^2 = D/\Sigma_a. \tag{7.224}$$

One-group recipes for obtaining M^2 from a few group constants D_n and Σ_{rn} are devised in ref. 5—for example, eq. (7.224) is replaced by

$$M^2 = \sum_n D_n/\Sigma_{rn}. \tag{7.225}$$

In LWRs M^2 varies between 50 cm^2 and 100 cm^2, depending mostly on the void fraction within the homogenized volume. The earlier introduction of k_∞ corresponds to an assumption of an infinite array of fuel assemblies. If the one-group fission yield cross-section is denoted by $\nu\Sigma_f$ then

$$k_\infty = \nu\Sigma_f/\Sigma_a. \tag{7.226}$$

Discretizing eq. (7.223) in one dimension yields readily

$$-2\phi_0 + \sum_{i=1}^{2}\phi_i + \frac{(k_{\infty 0} - 1)}{M^2}h_{i0}^2\phi_0 = 0 \tag{7.227}$$

where h_{i0} denotes the mesh spacing and the sum comprises the two neighbouring nodes. Since W_{i0} is equal for both values of i one has

$$W_{00} = 1 - 2W_{i0}. \tag{7.228}$$

Writing eq. (7.227) in a form analogous to eq. (7.22)

$$\phi_0 = \frac{k_{\infty 0} W_{i0} \sum_{i=1}^{2} \phi_i}{1 - k_{\infty 0} W_{00}} \tag{7.229}$$

it is observed that

$$W_{i0} = \frac{M^2}{h_{i0}^2 k_{\infty 0}}. \tag{7.230}$$

The FLARE kernel W_{i0} includes the expression in eq. (7.230) weighted by a factor g, where a lower order term $\sqrt{M^2}/h$ is given the weight $(1-g)$. The latter term can be inferred to prevail in the case of neutron transport from a slab of thickness h to a parallel contiguous and identical slab.

Reiterating the statements in a formal manner an expression of W_{i0}'s in eq. (7.222) is given by[44]

$$W_{i0} = (1-g)\frac{\sqrt{M_i^2}}{2h_{i0}} + g\frac{M_i^2}{h_{i0}^2} \tag{7.231}$$

where the mixing parameter is given as input. g posesses no clear physical meaning, but setting $g = 1$ would imply the finite difference coupling of eq. (7.227). As such, eq. (7.231) can be manifestly unphysical, viz. there is no requirement for $W_{00} = 1 - \Sigma W_{i0}$ to be non-negative. For this and a number of other reasons, more sophisticated techniques have been devised for deriving the pertinent physical expressions of the W_{i0}'s.[44]

Another perplexing feature in the FLARE scheme encompasses the treatment of the core–reflector interfaces where an albedo-like condition is applied. If the ith node is located at the core boundary, the leakage L_i is computed again via the transfer kernels W_{i0}. If the exterior material were black to neutrons then $L_i = n_i W_{i0} Q_i$ where n_i denotes the number of the adjacent nodes j which lie in the surrounding exterior volume. If the fraction α_i of neutrons is reflected black into the core, i.e. if α_i is the albedo of the node i defined in section 2.5, the leakage expression is modified to

$$L_i = W_{i0} Q_i (n_i - \beta_i) \tag{7.232}$$

where β_i is a modified albedo related to α_i. The proper α_i's cannot be applied directly since the mesh spacing h_{i0} included in the definition (7.231) is not meaningful at the core surface. If h_{is} denotes the half-thickness of the surface node i and if the surface transfer kernel W_{is} is defined by letting h_{is} replace h_{i0} then the two leakage expressions require[46]

$$(1-\beta_i)W_{i0} = (1-\alpha_i)W_{is}. \tag{7.233}$$

When the core-averaged albedo varies around 0.7 in LWRs the corresponding modified FLARE albedo is around 0.4.[46] Note that h_{is} is roughly $h_{i0}/2$.

The nodal calculation is usually tuned by means of adjusting the albedo β such that fine mesh results are reproduced as closely as possible. Again no guarantee can be given with regard to adjusted albedo values to be non-negative.

Feedback Effects

No matter which technique is chosen for the description of neutron diffusion, it must always be coupled to auxiliary calculations of feedback effects and fuel depletion as already shown in Fig. 7.13.

The axial feedback effects cause a considerable difference between BWRs and PWRs as far as core simulation is concerned. It has been pointed out already that the axial void fraction distribution is of indispensable importance in BWRs, while it is of lesser significance in PWRs. The PWR feedback effects on the axial power distribution are shown in Fig. 7.16. The graphs in Fig. 7.16 are taken from ref. 37 and they are calculated for the peak radial assembly. Note that the effects are mainly dependent on the power level.

In view of Fig. 7.16 it is clear that ignoring the axial effects even in a large PWR would cause the simulation model to deviate from reality. The discrepancies in the neutron flux distribution will exert their influence in the integral parameters as well. A simplified axial form of the flux distribution is shown in Fig. 7.17 separately in the cases of a true three-dimensional calculation and of a two-dimensional, i.e. zero axial buckling, approximation where the power distribution is axially constant.

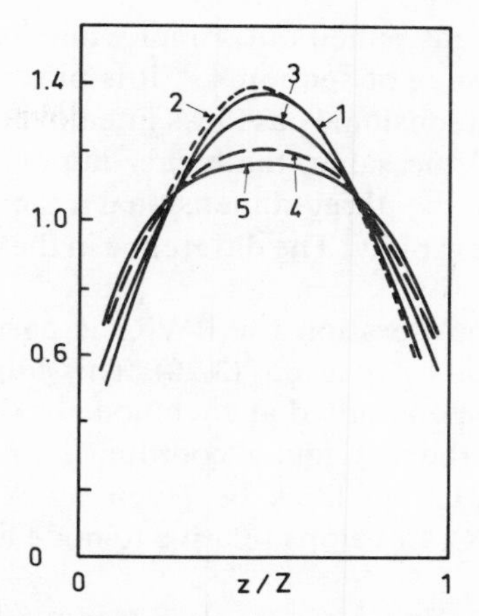

FIG. 7.16. Axial PWR power distribution with and without feedback effects for peak radial assembly. (1) No feedback effects, (2) hydraulic only, (3) hydraulic and xenon, (4) hydraulic and Doppler, (5) all feedback effects.[37]

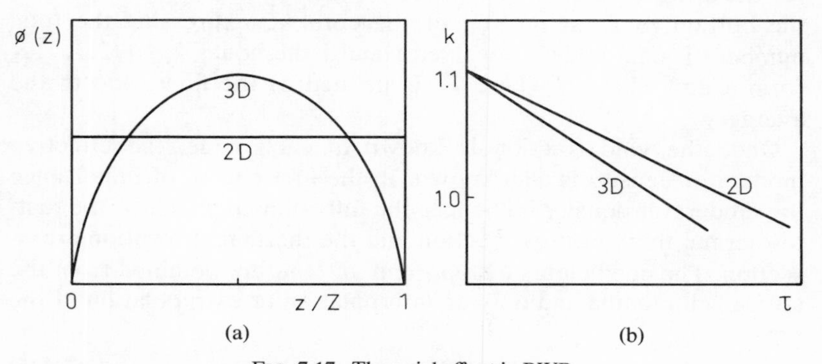

FIG. 7.17. The axial effect in PWR.

Since in both two- and three-dimensional calculation the flux integral $\int_0^z \phi dz$ is set to correspond to the same reactor power it is obvious that the three-dimensional flux shape exceeds the two-dimensional one in the central region of the core. This is, of course, assuming that the flux shape is convex, which is mostly the case in a core with control rods pulled out. Employing the physical argument of neutron importance of section 4.7, it is observed that the core is burnt in the two-dimensional case less intensively. Although the flux integral is one and the same, the higher importance of neutrons at the centre causes the three-dimensional reactivity curve in Fig. 7.17 (b) to fall more rapidly. The difference in the slopes is reported to be of the order of 5%.[37]

The void fraction iteration for BWRs is carried out in order to determine the value of α in eq. (3.20). This implies that the steam quality X_{ijk} must be computed at each node of the core. The indices i, j and k refer to the x, y and z coordinates, respectively. Let the core inlet and outlet qualities be given by X_i and X_o, then the nodewise quality X_{ijk} develops relative to node heat sources Q_{ijk} and is given by[44]

$$X_{ijl} = \frac{1}{KF_{ij}} \sum_{k=1}^{l} \left(Q_{ijk} - \frac{1}{2} Q_{ijl} \right) (X_0 - X_i) + X_i \qquad (7.234)$$

$$\text{for } l = 1, 2, \ldots, K$$

for the upward coolant flow where position index l varies from 1 at the bottom to K at the top of the core, K expressing the total number of axial nodes. For fixed i and j the nodes $k = 1, 2, \ldots, K$ form a flow channel where F_{ij} is defined as the flow ratio to the average.

Once the void fraction is known in each node, the effective moderator density is also known. In the first degree of importance the moderator density influences the diffusion coefficients, the fast-to-thermal transfer cross-section and the thermal absorption cross-section. The coefficients $\partial \Sigma / \partial \rho_M$ and $\partial D / \partial \rho_M$ are obtained from the lattice cell module and only an interpolation or extrapolation of the type

$$\Delta \Sigma = \frac{\partial \Sigma}{\partial \rho_M} (\rho_M - \rho_{M\,\text{nom}}) \,, \qquad (7.235)$$

$$\Delta D = \frac{\partial D}{\partial \rho_M} (\rho_M - \rho_{M\,\text{nom}}) \qquad (7.236)$$

is needed to update the cross-sections. ρ_M and $\rho_{M\,\text{nom}}$ refer to the real and nominal moderator densities. The derivatives are computed at $\rho_{M\,\text{nom}}$.

Among the feedback effects of nuclear origin, consider first the xenon effect. At this point one is reminded of the fact that the core simulators are not usually designed to reproduce violent distortions very exactly. As far as the xenon feedback is concerned, this means that only minor deviations from the equilibrium condition are included in the simulator capacity, whereas xenon oscillations are treated by separate routines which are linked to the core simulation programmes. The equilibrium xenon concentration N^{Xe} behaves as[5]

$$N^{Xe} = \frac{(y_1 + y_{Xe})\Sigma_f \phi_2}{\lambda_{Xe} + \sigma_a^{Xe} \phi_2} \tag{7.237}$$

and therefore the first order corrections due to xenon buildup are of the form

$$\Delta\Sigma_{a2} = \sigma_a^{Xe}(N^{Xe} - N^{Xe}_{nom}) \tag{7.238}$$

where N^{Xe}_{nom} refers to the equilibrium reference state. It is just the thermal absorption cross-section which is altered. Note that ϕ_2 in eq. (7.237) refers to the thermal flux.

The fission yield parameters y change with fuel composition and therefore with irradiation. Another noticeable poison isotope Sm^{149} is treated in a corresponding manner.

For the fuel temperature feedback it has already been indicated in section 4.2 that a $\sqrt{T}$ dependence is prevailing. In the two-group global LWR analysis the Doppler effect pertains to the fast group, where the coefficients $\partial\Sigma_{a1}/\partial\sqrt{T}$ are computed for the homogenized assembly compositions. The effect is described by

$$\Delta\Sigma_{a1} = \frac{\partial\Sigma_{a1}}{\partial\sqrt{T}}(\sqrt{T} - \sqrt{T_{nom}}) \tag{7.239}$$

where T_{nom} refers again to the nominal parameter combination for which the tabulation was done. As a matter of fact, modern algorithms devised for feedback effects construct interpolation splines of the forms

$$\Delta\Sigma_{a1}(x) = \sum_i a_i x^i \tag{7.240}$$

where

$$x = \frac{T - T_{nom}}{T_{nom}} \tag{7.241}$$

and a_i's are coefficients to be obtained by regression from the lattice cell results. The transfer cross-section Σ_{12} experiences a change of the same magnitude, but of a reversed sign as compared with the fast absorption cross-section.

1.5-Group Schemes

The requirement of minimizing the computer running costs is in vehement contrast to the need of two-group core analysis. If a one-group approach were used then the feedback effects would necessitate proper group constant averaging in order to recognize the fact that certain effects such as xenon are relevant for the thermal flux, whereas fuel temperature effects concern mainly the fast group energy domain. To satisfy the converse needs one may proceed by the one-group solution of the fast neutron flux combined with an introduction of a flux ratio[36, 41]

$$r_i = \phi_{2i}/\phi_{1i} \tag{7.242}$$

at each node i. Now i runs over the entire range of the three-dimensional nodes. The ultimate reason why the fast group is solved by diffusion methods can be found by considering the diffusion properties. Recalling that the fast diffusion length is longer than the thermal one, it is obvious that the coupling between the coarse mesh nodes is due to the fast streaming and the bulk of thermal neutrons are absorbed within the node by which they were slowed down. The introduction of a nodewise varying flux ratio accounts for the thermal leakage which is smaller but not necessarily negligible.

Ignoring the fission neutrons, the diffusion equation for the thermal group becomes, from eqs. (7.205) and (7.207),

$$-\nabla \cdot D_2 \nabla \phi_2 + \Sigma_{a2}\phi_2 = \Sigma_{s12}\phi_1. \tag{7.243}$$

Upon integrating eq. (7.243) conventionally over the node 0 yields

$$\frac{1}{h}\sum_{i=1}^{6} j_{2io} + \Sigma_{a2}^{0}\phi_{20} = \Sigma_{s12}^{0}\phi_{10} \tag{7.244}$$

where j_{2io} denotes the thermal neutron current across the interface between the nodes i and 0. Equation (7.244) is an exact variant of eq.

(7.208), the only differences being the inclusion of the energy group indices and the assumption of a uniform mesh where vertical and horizontal spacings are identical. The latter assumption is made just for simplicity.

Equation (7.244) yields for the flux ratio

$$r_0 = \frac{\Sigma^0_{s\,12}}{\Sigma^0_{a2}} - \frac{\sum_{i=1}^{6} j_{2io}}{h\Sigma^0_{a2}\phi_{10}} \tag{7.245}$$

where the thermal currents j_{2io} have to be eliminated.

Simple flux models can be employed to derive expression for j_{2io}. Assume, for example, that the system consists of two adjacent semi-infinite media, one having the group constants of the node 0 and the other possessing the properties of the node i. If the origin is located at the interface, then the one-dimensional solution of eq. (7.243) can be cast in the form

$$\phi_2 = A\ \mathrm{e}^{x/L_2} + B\ \mathrm{e}^{-x/L_2} + \frac{\Sigma_{s\,12}}{\Sigma_{a2}}\phi_1 \tag{7.246}$$

with $L_2 = D_2/\Sigma_{a2}$. Requiring the flux to be bounded causes the coefficient A to vanish for the half-space $x > 0$ and B to vanish for the half-space $x < 0$. The remaining coefficients, i.e. A for $x < 0$ and B for $x > 0$, are determined by the flux and current continuity across the interface. Solving for the current one obtains

$$J_{2io} = \frac{\Sigma^i_{s\,12}/\Sigma^i_{a2}\phi_{1i} - \Sigma^0_{s\,12}/\Sigma^0_{a2}\phi_{10}}{\dfrac{L_{2i}}{D_{2i}} + \dfrac{L_{20}}{D_{20}}}. \tag{7.247}$$

J_{2io} from eq. (7.247) is substituted in eq. (7.245) and the sum is extended over all the neighbouring nodes. Appropriate estimates for the fast fluxes must be used, since eq. (7.247) is not self-sufficient. The diffusion equation for the fast flux is given by

$$-\nabla\cdot D_1\nabla\phi_1 + \Sigma_{r1}\phi_1 = \lambda(\nu\Sigma_{f1} + r\nu\Sigma_{f2})\phi_1 \tag{7.248}$$

where the thermal flux appears disguised, typical of the 1.5 group approach. In particular x_1 is set equal to unity in eq. (7.248).

Observing that the flux ratio r depends implicitly on the fast flux ϕ_1 it is only natural that r can be updated at times when more accurate values of ϕ_1 become available. ϕ_{1i}'s will be solved iteratively and therefore one has the opportunity to improve r along with iterations. The derivation of the flux ratio was based on a crude technique, whereupon improvements can be found.[36]

7.6 Numerical Iteration Techniques

At various stages of the previous discussion situations were encountered where one has to employ iterative numerical techniques. The iteration compounds two separate steps: one concerned with the reduction of the multigroup eigenvalue problem to a series of coupled one-group problems and the other concerned with the numerical solution of the discretized one-group equations. The first step is referred to as the outer iteration and the second one is generally known as the inner iteration. The solution of eq. (7.125) is found by numerical algorithms pertinent to the inner iterations. On the other hand, eq. (7.179) involves both stages. For a fixed value of n the inner iteration is performed with the subsequent iterates and the estimation of λ being the outer iteration step. In section 7.5 the inner iterations concern the solution of eq. (7.208).

Outer Iteration

In order to demonstrate the idea of the outer iterations the multigroup diffusion equations (2.52) are expressed in a matrix notation by

$$(\mathbf{B}-\mathbf{S})\boldsymbol{\phi} = \lambda \mathbf{F}\boldsymbol{\phi} \tag{7.249}$$

where $\mathbf{B}$ is a diagonal matrix with elements

$$[\mathbf{B}]_{nn} = B = -\nabla \cdot D_n \nabla + \Sigma_{rn} \tag{7.250}$$

and the scattering matrix $\mathbf{S}$ possesses elements Σ_{smn}, while for the fission matrix one has

$$[\mathbf{F}]_{mn} = \chi_n (\nu\Sigma_f)_m. \tag{7.251}$$

The flux vector $\boldsymbol{\phi}$ has the group fluxes ϕ_n as its components.

The outer iteration scheme can be formulated as follows.[47] Let the fission source iterate $\boldsymbol{\psi}^{(i-1)}$ be known either from the previous iteration step or for $i = 1$ as the initial guess. Observe that $\boldsymbol{\psi}^{(0)}$ can be an arbitrary non-zero vector. The corresponding neutron flux distribution $\boldsymbol{\phi}^{(i)}$ is obtained from

$$(\mathbf{B}-\mathbf{S})\boldsymbol{\phi}^{(i)} = \boldsymbol{\psi}^{(i-1)}. \tag{7.252}$$

The solution of eq. (7.252) is precisely what is meant by the inner iteration. For the moment it is assumed that eq. (7.252) is solved and therefore $\boldsymbol{\phi}^{(i)}$ is known.

In terms of the fission source $\boldsymbol{\psi}$, eq. (7.249) is rewritten as

$$\mathbf{M}\boldsymbol{\psi} = k\boldsymbol{\psi} \tag{7.253}$$

where the matrix $\mathbf{M}$ is seen to be $\mathbf{M} = \mathbf{F}(\mathbf{B}-\mathbf{S})^{-1}$ and k is the effective multiplicative factor $k = 1/\lambda$. The general iteration step is expressed now by

$$\boldsymbol{\psi}^{(i)} = \mathbf{M}\boldsymbol{\psi}^{(i-1)} = \mathbf{M}^{i}\boldsymbol{\psi}^{(0)}. \tag{7.254}$$

To demonstrate that the scheme converges to the dominant eigenvalue $\boldsymbol{\psi}^{(0)}$ is expanded in terms of the eigenvectors $\mathbf{x}^{l}$ of $\mathbf{M}$[47],

$$\boldsymbol{\psi}^{(0)} = \sum_{l} a_l \mathbf{x}^{l}. \tag{7.255}$$

$\boldsymbol{\psi}^{(i)}$ is now expressed by

$$\boldsymbol{\psi}^{(i)} = \sum_{l} a_l M^{i}\mathbf{x}^{l} = k_0^{i} \sum_{l} a_l \left(\frac{k_l}{k_0}\right)^{i} \mathbf{x}^{l} \tag{7.256}$$

where k_0 denotes the dominant value of k. When $i \to \infty$ and $k < k_0$, one obtains

$$\boldsymbol{\psi}^{(i)} \xrightarrow[i\to\infty]{} k_o^{i} a_0 \mathbf{x}^{0}. \tag{7.257}$$

There are a number of procedures to solve the eigenvalue estimate k_0. A possibility is simply

$$k_0^{(i)} = \frac{(\boldsymbol{\psi}^{(i-1)}, \boldsymbol{\psi}^{(i)})}{(\boldsymbol{\psi}^{(i-1)}, \boldsymbol{\psi}^{(i-1)})} \tag{7.258}$$

where the vector inner product is the integration over the reactor volume V, i.e.

$$(\boldsymbol{\phi}, \boldsymbol{\psi}) = \sum_{n=1}^{N} \int_{V} \phi_n(\mathbf{r})\psi_n(\mathbf{r})\, d\mathbf{r}, \tag{7.259}$$

the sum being extended over all the N energy groups.

Since a direct application of eq. (7.257) would cause numerical difficulties when k_0 deviates substantially from unity, the new fission source iterate $\boldsymbol{\psi}^{(i)}$ is normalized at each step dividing it by $k_0^{(i)}$. The vector

$$\boldsymbol{\psi}^{(i)} \longrightarrow \frac{1}{k_0^{(i)}} \boldsymbol{\psi}^{(i)} \tag{7.260}$$

is hence employed in eq. (7.252) for $i = i + 1$. The subscript 0 is omitted elsewhere in the present discussion where the domination of k_0 is less emphasized.

The outer iterations can be performed in alternative ways,[47] but the power (or source) iteration method outlined above is used most frequently. The convergence can be accelerated appreciably by adopting a technique where the source iterate $\boldsymbol{\psi}^{(i)}$ is damped, introducing a linear combination of the previous iterates and using an expression

$$\boldsymbol{\psi}^{(i)} = \sum_{j=0}^{i} a_{ij} \boldsymbol{\chi}^{(j)} \tag{7.261}$$

instead of eq. (7.260). Proceeding along this line[47] one would select the coefficients a_{ij} in a manner that tends to minimize the error. For example, a_{ij}'s can be obtained by means of Chebyshev minimax theory and, in fact, the related scheme known as the Chebyshev polynomial extrapolation is used widely.[47]

Inner Iteration

Returning now to the solution of eq. (7.252) and recalling that the matrix $\mathbf{B}$ is a diagonal one, it is observed that the group-to-group coupling can be removed on the left-hand side if the inscattering source is included in the known iterate and the term $\mathbf{S}\boldsymbol{\phi}$ is taken over to the right by lowering its index of iteration,

$$\mathbf{B}\boldsymbol{\phi}^{(i)} = \mathbf{S}\boldsymbol{\phi}^{(i-1)} + \boldsymbol{\psi}^{(i-1)}. \tag{7.262}$$

Equation (7.262) is now a general expression combining all the previous cases where the inner iteration is required. The one group operator B_n and the equation

$$B_n \phi_n^{(i)} = q_n^{(i)} \tag{7.263}$$

are discretized pursuant to the discussion of sections 7.4 and 7.5. $q_n^{(i)}$ denotes the nth component of the source

$$q_n^{(i)} = [\mathbf{S}\boldsymbol{\phi}^{(i-1)} + \boldsymbol{\psi}^{(i-1)}]_n. \tag{7.264}$$

Once the inner iteration is separated from the outer one and the nth group from the rest there will be no need to use the indices i and n to remind one of the exterior problem.

After performing the discretization one is faced with a linear system of equations of which examples are furnished by eqs. (7.128), (7.179), (7.208) and (7.219). In general, the problem boils down to a linear transformation of the inhomogeneous type

$$\mathbf{A}\boldsymbol{\phi} = \mathbf{q} \tag{7.265}$$

where the components of $\boldsymbol{\phi}$ are now the pointwise flux values ϕ_{ijk} ordered in a one-dimensional array. The elements of **A** represent the coupling between the nodes and are typically of the form given in eqs. (7.180)–(7.184) or eqs. (7.209)–(7.224), for example. The matrix **A** is rather sparse in many cases so that a great number of elements are zero. As was pointed out in section 7.4, the collision probability method is one of the rare situations with a full matrix **A**.

In most cases general iterative methods devised for the system of equations of the type in eq. (7.265) are used for the solution. There are a few exceptions where due to a small number of nodes the rank of **A** is low enough to permit direct inversion. Ultimately the choice between the direct and iterative methods depends on the computing environment and computer hardware. The finite element method involves a rather convenient form for the matrix **A** and there the Cholesky factorization technique has become popular in the direct inversion.[39] However, the absolute majority of computer codes employ iterative methods in conjunction with other discretization techniques.

The standard iterative methods applicable to eq. (7.265) are discussed widely in the literature[30, 47] and therefore only a brief outline will be given below.

The matrix **A** is first split into three parts,

$$\mathbf{A} = \mathbf{D} - \mathbf{E} - \mathbf{F} \tag{7.266}$$

where $\mathbf{D}$ comprises the diagonal elements of $\mathbf{A}$ while $-\mathbf{E}$ and $-\mathbf{F}$ consist of the lower and upper diagonal parts of $\mathbf{A}$.

A straightforward technique to solve eqs. (7.265) by employing the splitting is to execute an iterative scheme

$$\mathbf{D}\boldsymbol{\phi}^{(i)} = (\mathbf{E}+\mathbf{F})\boldsymbol{\phi}^{(i-1)} + \mathbf{q} \tag{7.267}$$

or equivalently

$$\boldsymbol{\phi}^{(i)} = \mathbf{D}^{-1}(\mathbf{E}+\mathbf{F})\boldsymbol{\phi}^{(i-1)} + \mathbf{D}^{-1}\mathbf{q}. \tag{7.268}$$

Gauss's method used in eq. (7.268) is frequently called simultaneous relaxation.[47]

Another scheme referred to as successive relaxation[47] avoids the storing of all the components of the previous flux estimate. If eq. (7.267) is manipulated into

$$(\mathbf{D}-\mathbf{E})\boldsymbol{\phi}^{(i)} = \mathbf{F}\boldsymbol{\phi}^{(i-1)} + \mathbf{q} \tag{7.269}$$

the successive technique iterates as

$$\mathbf{q}^{(i)} = (\mathbf{D}-\mathbf{E})^{-1}\mathbf{F}\boldsymbol{\phi}^{(i-1)} + (\mathbf{D}-\mathbf{E})^{-1}\mathbf{q}. \tag{7.270}$$

Convergence of the iteration can be accelerated by introducing a parameter ω which combines $\boldsymbol{\phi}^{(i)}$ and $\boldsymbol{\phi}^{(i-1)}$ in the iteration process. From eq. (7.269) one has an auxiliary estimate $\hat{\boldsymbol{\phi}}^{(i)}$

$$\hat{\boldsymbol{\phi}}^{(i)} = \mathbf{D}^{-1}(\mathbf{E}\boldsymbol{\phi}^{(i)} + \mathbf{F}\boldsymbol{\phi}^{(i-1)}) + \mathbf{D}^{-1}\mathbf{q} \tag{7.271}$$

which is coupled to $\boldsymbol{\phi}^{(i-1)}$ by

$$\boldsymbol{\phi}^{(i)} = \omega\hat{\boldsymbol{\phi}}^{(i)} + (1-\omega)\boldsymbol{\phi}^{(i-1)}. \tag{7.272}$$

Substituting eq. (7.271) in eq. (7.272) yields after some manipulation

$$\boldsymbol{\phi}^{(i)} = \mathbf{M}\boldsymbol{\phi}^{(i-1)} + \mathbf{k} \tag{7.273}$$

where

$$\mathbf{M} = (\mathbf{D}-\omega\mathbf{E})^{-1}[(1-\omega)\mathbf{D}+\omega\mathbf{F}] \tag{7.274}$$

and

$$\mathbf{k} = \omega(\mathbf{D}-\omega\mathbf{E})^{-1}\mathbf{q}. \tag{7.275}$$

The utility of the parameter ω is seen qualitatively by considering the convergence properties of the scheme in eq. (7.273). Clearly the exact solution $\boldsymbol{\phi}$ satisfies

$$\boldsymbol{\phi} = \mathbf{M}\boldsymbol{\phi} + \mathbf{k} \tag{7.276}$$

and if the error at the iteration step i is denoted by $\boldsymbol{\epsilon}^{(i)}$ it is seen from eqs. (7.273) and (7.275) that

$$\boldsymbol{\epsilon}^{(i)} = \mathbf{M}\boldsymbol{\epsilon}^{(i-1)} = \mathbf{M}^i\boldsymbol{\epsilon}^{(0)} \tag{7.277}$$

where $\boldsymbol{\epsilon}^{(0)}$ is the error contained in the initial guess. Since $\boldsymbol{\epsilon}^{(0)}$ is non-zero $\boldsymbol{\epsilon}^{(i)}$ converges to zero if and only if $\mathbf{M}^i \rightarrow \mathbf{0}$ when i is increased. Selecting ω in an optimal manner would obviously cause the iteration to converge more rapidly. The most efficient convergence would be achieved by choosing[47]

$$\omega = \frac{2}{1+\sqrt{1-\mu^2}} \tag{7.278}$$

where μ denotes the spectral radius, i.e. the largest eigenvalue, of the matrix $\mathbf{D}^{-1}(\mathbf{E}+\mathbf{F})$. Equations (7.273)–(7.275) describe the successive over-relaxation method which has achieved the central importance in the core analysis codes. The optimum value of ω is seldom based on eq. (2.278), but less accurate estimates are for a numerical basis.

The practical execution of the successive over-relaxation method still involves a number of tricks which are not always understood with the profound theoretical rigour characteristic of the standard formulation. Mainly due to difficulties one may encounter in handling linear transformation, the matrix $\mathbf{A}$ is frequently considered in blocks. For example, the solution $\boldsymbol{\phi}$ of eq. (7.265) can be divided into two sets, the vector $\boldsymbol{\phi}_1$ being composed of one set and the vector $\boldsymbol{\phi}_2$ of the other. Introducing the corresponding blocks $\mathbf{A}_{ij}$ and dividing the source vector $\mathbf{q}$ into two parts, eq. (7.265) is partitioned as

$$\begin{pmatrix} \mathbf{A}_{11} & \mathbf{A}_{12} \\ \mathbf{A}_{21} & \mathbf{A}_{22} \end{pmatrix} \begin{pmatrix} \boldsymbol{\phi}_1 \\ \boldsymbol{\phi}_2 \end{pmatrix} = \begin{pmatrix} \mathbf{q}_1 \\ \mathbf{q}_2 \end{pmatrix} \tag{7.279}$$

and $\boldsymbol{\phi}_1$ can be solved iteratively commencing from

$$\mathbf{A}_{11}\boldsymbol{\phi}_1^{(i)} = \mathbf{q}_1^{(i)} - \mathbf{A}_{12}\boldsymbol{\phi}_2^{(i-1)} \tag{7.280}$$

where the previous iterates are used in the components of $\boldsymbol{\phi}_2$.

A concrete example of the scheme in eq. (7.280) is found by letting $\boldsymbol{\phi}_1$ consist of the set $\{\phi_{ij1}, \phi_{ij2}, \ldots, f_{ijk}\}$ representing all the axial nodes in a given flow channel specified by the fixed pair i and j. One would actually separate the flow channel from the rest of the reactor from which the coupling is placed in the known source term. According to the physical meaning of the matrix blocks, one can speak of point, line or even plane relaxation.

7.7. Computer Code Libraries

For a reactor engineer it is of great importance to have a fairly good and complete arsenal of computer codes available in fuel management work. In this case the completeness is closely equivalent to the capability of covering all the phases discussed earlier in this chapter. Excluding the reactor vendors and perhaps the largest power utilities, no self-sufficient own investment is conceivable in acquiring the analytical capability. Remarkable support is given then by the computer code centres established for the proliferation of digital computer programs and information pertaining thereto.

This section is included in the text for the purpose of outlining a system of the non-proprietary codes which are deposited either at the Argonne Code Center or at the NEA Computer Program Library. Although certain restrictions may be imposed on the code acquisitions, one has reason to acknowledge the level at which the libraries stand. No review of the commercially developed fuel management code systems will be undertaken.

Due to the fact that the code development had commenced earlier than the organized buildup of nuclear data compilations, the lattice codes usually include multigroup cross-section libraries of their own. However, if the user is willing to exploit the compiled original data, processing codes are available for calculating the MUFT and THERMOS libraries. The LASER code[24] is a combination of the two cell programs and covers the entire energy domain with a depletion routine incorporated. One should recall the earlier remarks on the rather long running times involved and evidently some effort would be worth while to modify the code before its installation for production work. In the LEOPARD code[48] the thermal energy spectrum is computed by a different method using the thermal disadvantage factor and without an explicit application of transport theory solutions. Both LASER and LEOPARD have a large progeny of proprietary versions. The British WIMS code[31] represents the flexible and more sophisticated cell burnup programs with a version listed in ref. 49.

For use in reactor simulation, later versions of FLARE[43] have been deposited in the libraries. In conjunction with the discussion on the nodal method, lenient criticism has been directed towards the algorithm which still has the salient property of involving rather low computer costs.

The selections mentioned above are only certain specific codes picked out from library information consisting of up to a few hundred program items. Consistently with the detailed description of the methods, the examples can be used to construct a code system based on the implicit burnup calculation. It would be misleading to ignore the codes where depletion is coupled explicitly to the global core analysis. Recalling that this approach will inevitably entail higher costs of computation and an accuracy that may easily exceed the utility needs, one should mention the PDQ system[50] with its multiple phases of development.

References

1. Panel summary, in *Reactor Burnup Physics*, International Atomic Energy Agency, Vienna, 1973.
2. Ozer, O. and Garber, D., ENDF/B Summary Documentation. ENDF-201, BNL 17541, Brookhaven National Laboratory, Upton, New York, 1973.
3. BWR 6, *General Description of a Boiling Water Reactor*, General Electric, San Jose, Calif., 1973.
4. Lim, E. Y. and Lenard, A., *Trans. Am. Nucl. Soc.* **19**, 172 (1974).
5. Zweifel, P. F., *Reactor Physics*, McGraw-Hill, New York, 1973.
6. Ferziger, J. H. and Zweifel, P. F., *The Theory of Neutron Slowing Down in Nuclear Reactors*, Pergamon Press, 1966.
7. McGoff, D. J., NAA-SR-Memo-5766, Atomics International, Canoga Park, California, 1960.
8. Siltanen, P., VTT-YDI-6, Technical Research Centre of Finland, Helsinki, 1973.
9. Bell, G. I. and Glasstone, S., *Nuclear Reactor Theory*, Van Nostrand Reinhold Company, New York, 1970.
10. Sauer, A., *Nucl. Sci. Engng.* **16**, 329 (1963).
11. Carlvik, I., *Nucl. Sci. Engng.* **29**, 325 (1967).
12. Ishiguro, Y., *Nucl. Sci. Engng.* **32**, 422 (1968).
13. Höglund, R. and Wasastjerna, F., VTT-YDI-16, Technical Research Centre of Finland, Helsinki, 1975.
14. Borresen, S. and Goldstein, T., *Trans. Am. Nucl. Soc.* **15**, 296 (1972).
15. Honeck, H. C., BNL-5826, Brookhaven National Laboratory, Upton, New York, 1961.

16. Beardwood, J. E. *et al.*, in ANL-7050, Argonne National Laboratory, Argonne, Illinois, 1965.
17. Carlvik, I., Report 506, Helsinki University of Technology, 1969.
18. Williams, M. M. R., *Mathematical Methods in Particle Transport Theory*, Butterworths, London, 1971.
19. Patrakka, E. and Saastamoinen, J., in *Numerical Reactor Calculations*, International Atomic Energy Agency, Vienna, 1972.
20. Cheng, H. S., in CONF-710302, U.S. Atomic Energy Commission, 1971.
21. Häggblom, H. *et al.*, *Nucl. Sci. Engng.* **56**, 411 (1975).
22. Karppinen, J. and Saastamoinen, J., *Trans. Am. Nucl.* **17**, 262 (1973).
23. Mayer, L., Bericht 4/68-21, Kernforschungszentrum, Karlsruhe, 1969.
24. Poncelet, C. F., WCAP-6073, Westinghouse Electric Corporation, Pittsburgh, Penn., 1966.
25. Kaikkonen, H., Diploma thesis, Helsinki University of Technology, 1975.
26. Bonalumi, R. A., *Trans. Am. Nucl. Soc.* **14**, 226 (1971).
27. Honeck, H. C. and Finch, D. R., DP-1278, Savannah River Laboratory, Aiken, South Carolina, 1971.
28. Williams, M. M. R., *The Slowing Down and Thermalization of Neutrons*, North-Holland Publishing Company, Amsterdam, 1966.
29. Esch, L. *et al.*, *Nucl. Sci. Engng.* **46**, 233 (1971).
30. Varga, R. S., *Matrix Iterative Analysis*, Prentice-Hall, Englewood Cliffs, New Jersey, 1962.
31. Fayers, F. J. *et al.*, AEEW-R-785, Atomic Energy Establishment, Winfrith, Dorchester, 1972.
32. Fredin, B. *et al.*, *Nucl. Sci. Engng.* **36**, 315 (1969).
33. Karppinen, J. and Saastamoinen, J., VTT-YDI-9, Technical Research Centre of Finland, Helsinki, 1973.
34. Crowther, R. L., in CONF-730414, U.S. Atomic Energy Commission, 1973.
35. England, T. R., WAPD-TM-333, Westinghouse Electric Corporation, Pittsburgh, Penn., 1962.
36. Borresen, S., *Nucl. Sci. Engng.* **44**, 37 (1971).
37. Fayers, F. J. and Nash, G., *Annals of Nucl. Sci. Engng.* **1**, 185 (1974).
38. Henry, A. E., in *Numerical Reactor Calculations*, International Atomic Energy Agency, Vienna, 1972.
39. Strang, G. and Fix, G. J., *An Analysis of the Finite Element Method*, Prentice-Hall, Englewood Cliffs, New Jersey, 1973.
40. Deppe, L. O. and Hansen, K. F., *Nucl. Sci. Engng.* **54**, 456 (1974).
41. Robinson, C. P. and Eckard, J. D., TIS-3351, Combustion Engineering Inc., Windsor, Conn., 1972.
42. DOCKET-RESARA-16, Westinghouse Nuclear Steam Supply System, U.S. Atomic Energy Commission, Technical Information Center, Oak Ridge, Tenn., 1973.
43. Henry, A., *Nuclear-Reactor Analysis*, MIT Press, Cambridge, Mass., 1975.
44. Delp, D. L. *et al.*, GEAP-4598, General Electric Co., San Jose, Calif., 1964.
45. Maeder, C. and Varandi, G., EIR-234, Eidg. Institut für Reaktorforschung, Würenlingen, 1973.
46. Shimooke, T. and Mochizuki, K., in *The Physics Problems in Thermal Reactor Design*, The British Nuclear Energy Society, London, 1967.

47. Wachspress, E. L., *Iterative Solutions of Elliptic Systems*, Prentice-Hall, Englewood Cliffs, New Jersey, 1966.
48. Strawbridge, L. E. and Barry, R. F., *Nucl. Sci. Engng.* **23**, 58 (1965).
49. News from CPL, OECD-NEA Computer Programme Library, Ispra, 1975.
50. Cadwell, W. R., WAPD-TM-678, Westinghouse Electric Corporation, Pittsburgh, Penn., 1967.

CHAPTER 8

Alternative Reactor Concepts

THE reactor core design exerts its influence to fuel management in a number of ways discussed earlier. An assessment of the advantages of a given design must include the viewpoint of core management. Since fuel costs are less than one-third of the power generating costs, it is clear that the reactor core fuel management aspects cannot alone dictate the competitiveness.

To recognize the wide interest devoted to the designs other than the LWR, a brief summary will be given on the other reactor types where industrial commitments have been made. Coherently with the main topic, only core analysis aspects will be discussed with a particular emphasis on the differences from uranium-fuelled LWRs. To a large extent the core analysis systems constructed for these reactors follow the idea of what is presented in Chapter 7 for LWRs. Therefore no quantitative development is necessary.

8.1. Plutonium Recycle in LWRs

A 1000 MWe BWR produces typically some 150 kg of fissile plutonium per year while the figure is about 175 kg for a corresponding PWR. The plutonium credit is of the order of 10% of the fuel costs. The potential capability of LWRs to use plutonium as a primary fuel has created large demonstration programs and there is certainly an economic incentive to harness some of the plutonium in LWRs rather than wait for fast breeders to do it.

The value of plutonium in thermal reactors can generally be based on the worth estimated for highly enriched uranium.[1] If C_U denotes the cost of 93% enriched uranium, then the value of plutonium C_{Pu}

can be expressed by[1]

$$C_{Pu} = AC_U - BC_0 \tag{8.1}$$

where A denotes the relative worth of Pu^{239} to U^{235} and is equal to about 0.9. In the correction term C_0 is the ratio of Pu^{242} content to $Pu^{239} + Pu^{241}$ content in plutonium and the constant $B = 0.16$ is reported in ref. 1. Due to the varying modes of plutonium exploitation C_0 is expected to increase until the mid-1980s when part of the discharged plutonium comes from recycling.

In conjunction with reactor core fuel management, plutonium recycling is seen as a problem of neutronics. Firstly, the cross-sections are known less accurately. Relative to uranium cores large amounts of plutonium in the core will incorporate a complicated resonance structure in addition to the fact that more isotopes will be present. Performing reactor physics calculations by standard routines will not therefore yield the normal precision. Secondly, fewer degrees of freedom will be available in the fuel assembly design and in planning the reload batches. This is because one would desire to maintain the same safety and performance limits regardless of whether the reload is made of uranium or mixed plutonium–uranium fuel.[2]

Plutonium-fuelled lattices have a harder, i.e. more energetic, neutron spectrum relative to uranium lattices. Control devices whether in control rods or in a soluble form include thermal absorbers and consequently the reaction rates in the absorbing materials are reduced. In other words, the worth of control rods and soluble poisons as well as thermally absorbing fission products, e.g. xenon, is reduced. This is an important aspect in view of the requirement that the rod worths would be at least the same as in uranium-fuelled lattices.[3]

Examining the control requirements one observes a reverse trend. Due to increased resonance effects the Doppler coefficient becomes more negative with increasing plutonium inventory. For the same reason and due to the spectrum shift moderator temperature coefficient becomes more negative as well. In order to maintain the same manoeuvrability the control requirement is increased from what one has in a uranium lattice. To this effect there are some contributors as well, viz. the smaller delayed neutron fraction implies a slightly less stable system and even local power peaking increases near the water gaps.[3]

In BWR lattices where the control rods are located in the water gaps between fuel assemblies a natural method to alleviate the above difficulties has been to develop an island concept.[4] The highly enriched plutonium, up to 4%, is located in the central parts of the assembly. There is a concomitant benefit because the interior uranium assembly already had a higher enrichment than average (see Fig. 7.3), and therefore savings are gained in enrichment costs.

A minor penalty is incurred in PWRs where instead of a uniform enrichment the plutonium assemblies will have at least three degrees of enrichment.[3] The enrichment pattern depends on the positions of the control rod clusters. Besides the discrete assembly concept with all-plutonium fuel pins a graded enrichment concept has also been explored. There one has both uranium and plutonium pins within a single assembly.[3]

In order to illustrate the local power peaking induced by water slots in plutonium-fuelled systems, consider Fig. 8.1 where a one-

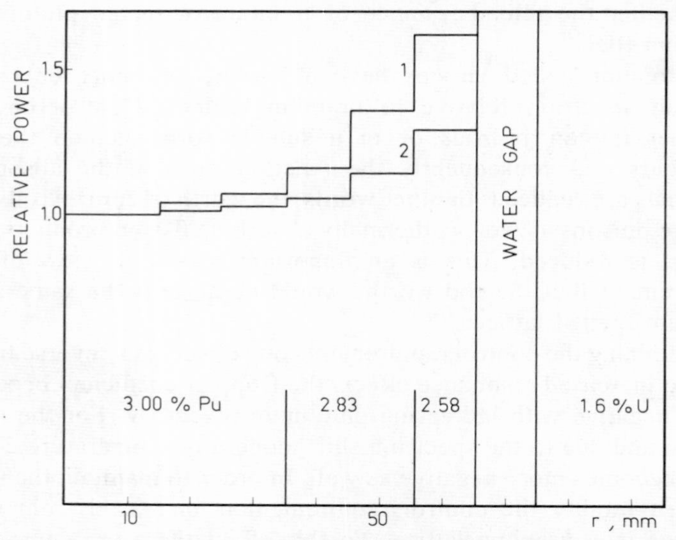

FIG. 8.1. Comparison of power peaking within plutonium (1) and uranium (2) fuel assemblies.

dimensional cut of a heterogeneous lattice is depicted. Within the discrete plutonium assembly the enrichment varies from 3.02% to 2.58%, decreasing towards the assembly boundary. The adjacent 1.6% uranium assembly is separated by a narrow water gap. It is seen in Fig. 8.1 how the relative power distribution peaks towards the water gap. In comparison with 3.6% enriched uranium the plutonium assembly yields a power peak of some 25% greater.

8.2. HTGR

The high temperature gas-cooled reactor differs substantially from LWRs using a different fuel, coolant and moderator. At the present stage the HTGR is charged by highly enriched (93%) U^{235} with large amounts of Th^{232} being loaded in the core as fertile material. Th^{232} is converted into U^{233} and the HTGR core possesses an immediate capability to operate on the thorium–uranium recycle mode with some fully enriched uranium as the makeup feed material. A low enriched cycle based on U^{235} is feasible as well.

The HTGR core design appears in nuclear engineering journals and in reports issued by the vendor.[5] The fuel element is a hexagonal graphite block with holes for the fuel particles and the coolant. Helium is used as the coolant and the graphite serves as the moderator. The fuel particles are all ceramic. The fuel elements are ordered in columns. Each column has eight fuel elements and the reactor core is radially composed of some 500 columns. The reactor size is scaled up by increasing the number of fuel columns.[5]

The standard core management scheme employs annual reloading where one-fourth of the fuel is discharged at each refuelling. The core is divided into radial zones. To reduce the global radial peaking factor the outer zones have a higher uranium concentration relative to the inner zones where thorium is more heavily loaded. The helium flow and heat transfer characteristics favour axial zoning. The number of axial zones is two or three. To yield an axially constant fuel temperature the flux shape is allowed to peak towards the coolant inlet.[5]

In the core analysis one has to consider the dualism between the thorium and uranium contributions. Due to the cross-section structure of Th^{232} and resonance effects the fast neutron spectrum is linked to the C/Th ratio, whereas the thermal domain is characterized by the carbon-to-fissile uranium ratio. The spectrum calculations proceed as in section 7.3. The coated fuel particles are bonded into fuel rods and the cell geometry becomes more heterogeneous than in the LWR. A cell is schematically represented in Fig. 8.2.

The thermal cutoff energy is usually placed at 2.38 eV. The collision with graphite may well upscatter neutrons above this energy and therefore there would be a complicated coupling with the fast and the thermal groups. For this reason four groups are frequently used in the HTGR analysis. The lowest fast group is narrow enough to include only the energies below the lowest Th^{232} resonance at 21.9 eV. If all neutrons with energies above 2.38 eV were treated as fast fission neutrons, severe errors would be involved.[6] In the static reactor calculations the diffusion equation is discretized in hexagonal configuration by applying the method of integration over the nodes. In case the nodes obey the same symmetry as the lattice, then the radial node cross-section is a hexagon and the node possesses six adjacent neighbours in addition to the two vertical ones. The discretized form [cf. eq. (7.208)] is

$$C_{00}\phi_0 - \sum_{i=1}^{8} C_{i0}\phi_i + \Sigma_{a0}\phi_0 = q_0 \tag{8.2}$$

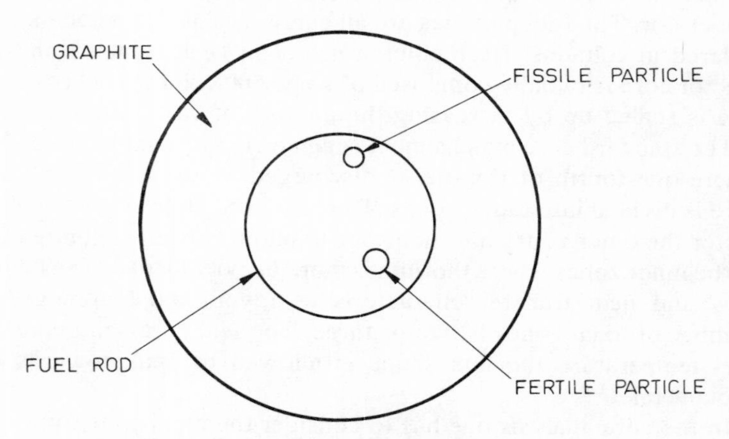

FIG. 8.2. HTGR cell geometry.[5]

where the six radial coupling coefficients are given by

$$C_{io} = \frac{4}{3h^2} \frac{D_i D_o}{D_i + D_o} \tag{8.3}$$

while the vertical ones remain unchanged and obey eq. (7.210). Despite the increased number of the nearest neighbouring nodes eq. (8.2) is solved numerically by the techniques discussed in section 7.6.

8.3. SGHWR

After one 100 MWe prototype operating from the 1960s the steam generating heavy water reactor has gained prospects of becoming commercial in the United Kingdom by the end of the 1970s. The scaling up of a reactor power is relatively easy from the core design point of view, where it mainly involves only increasing the number of pressure tubes.

In the SGHWR the fuel rods are clustered within a pressure tube

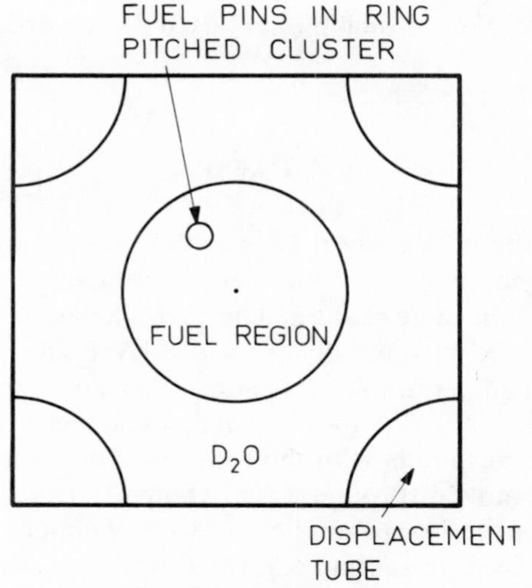

FIG. 8.3. SGHWR lattice cell.

to form water-coolant fuel channels. There are 36 fuel pins per cluster and the number of fuel channels is above 500 in a 660 MWe core.[7] The tubes are inserted in calandria filled with the heavy water moderator. As is indicated in Fig. 8.3, the lattice includes displacement tubes which are either empty or filled with the bulk moderator. Lattice cell homogenization is performed in a manner similar to the method discussed in section 7.3. The cylindrical equivalence cell is divided in a number of regions characterizing the neutron spectra. Instead of the lengthy MUFT–THERMOS procedures, the British have devised a fast-running simplified method incorporated in the METHUSELAH code.[9] The pressure tube version of METHUSELAH has three fast groups above 0.625 eV. The thermal region is covered by two overlapping groups, one representing the heavily absorbing fuel cluster and the other representing the moderator where absorption is low. For the five-group equations a cross-section library is established by a previous cell analysis. The two thermal groups are coupled via scattering. The simplified scheme is attractive in reactor core fuel management where the computations are fast. In fact, METHUSELAH has a four-group version for LWRs as well.

Overall core analysis of the SGHWR is conducted within a two-group three-dimensional model based on the ordinary diffusion theory approach.[8]

8.4. CANDU

CANDU is the only reactor type with a natural uranium-fuelled core that has maintained its commercial competitiveness after the nuclear industry became mature. The fuel management of CANDU differs even more substantially from the LWR than the SGHWR.

The established terminology is somewhat confusing since in the CANDU nomenclature[10] the fuel pellets are sheathed to a fuel element that corresponds with the rod. The elements are ordered in a bundle to be moved through a fuel channel of the reactor. There are axially about a dozen bundles in each channel.

As was discussed in section 6.3, the CANDU design exhibits the capability for on-power refuelling which is unique among commercial reactors. The small excess reactivity involved necessitates this fuelling arrangement. Another particular facet in CANDU is the axially bidirectional fuelling where the core axis is horizontal and the core is symmetrical with respect to the midplane.[11]

Core analysis consists again of two phases: the calculation of lattice parameters and the three-dimensional reactor simulation. In principle, there exists no obstruction to use the general cell program scheme outlined in section 7.4, but the Canadians have developed a simpler approach[11] which satisfies the computational need. No explicit transport or diffusion theory formulates are used. The processes are combined in terms of the four basic considerations: the fuel efficiency η, fast fission ϵ, resonance escape p and thermal utilization f. The leakage is treated by two-group bucklings. Diverse semi-empirical relations are then used to deduce the spectrum characteristics. There is an option to derive the lattice parameters either as a function of exposure or homogenized as time-averaged.[11]

In heavy water lattices the neutron migration area is very large and coarse meshes have no severe physical implications. Mesh intervals up to 0.57 m are reportedly[11] in regular use. Full-core three-dimensional calculations are performed in two energy groups using the normal finite diffusion techniques.

8.5. WWER

The WWER[12] designed in the Soviet Union is a PWR with some modifications compared with the descriptions given earlier. A 1000 MWe version is expected to operate in the late 1970s. The reactor physics of the WWER reactors does not warrant further discussion, but Chapter 7 is directly applicable, with the exception that the core has a hexagonal configuration. Therefore, the diffusion routines, for example, must be tailored according to eqs. (8.2)–(8.3).

Attention may be paid to the fact that the reload enrichment is a few tenths of 1% higher than what one is accustomed to. The 1000 MWe version employs an enrichment of 4.4%. The difference is evidently due to the fact that the core diameter is slightly smaller, which enhances leakage.

The smaller WWER-440 reactor has an interesting fuel management aspect attached to it. Partially due to the control rod structure there are usually two enrichments present in the reloading batch. The enrichments are prespecified. If the desired enrichment varies from cycle to cycle, the relative size of the two differently enriched batch fractions is altered. Figure 8.4 depicts an equilibrium cycle loading pattern.[13] Two different patterns alternate in the scheme of Fig. 8.4. This is because the central assembly is a batch of its own and is discharged after every two years.

Having the initial fuel enrichment specified *a priori* reduces the degrees of freedom. A conceivable tradeoff is achieved by shortening the lead time required for the issuance of the reload enrichment specification. This is naturally the case only if there are other similar reactors operating with the same reload enrichment.

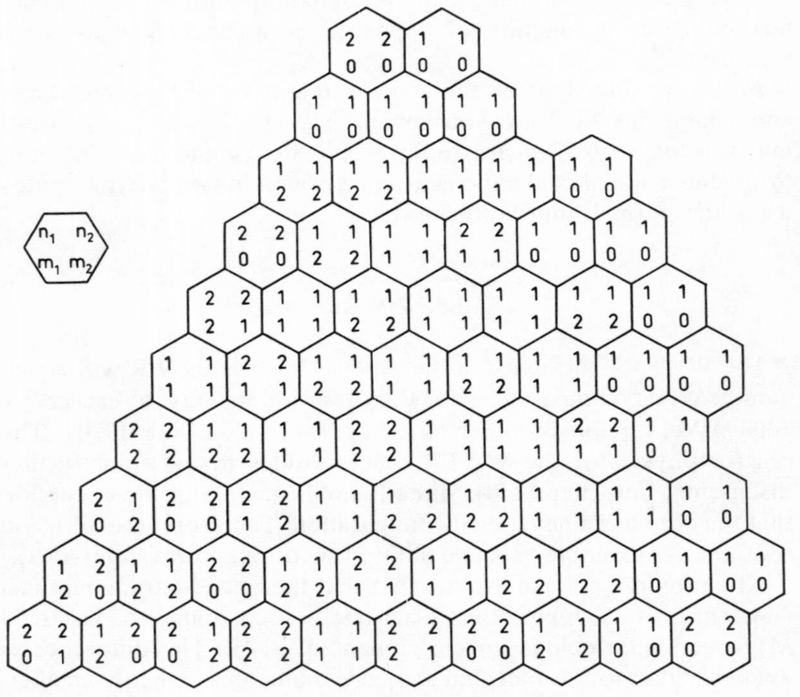

FIG. 8.4. WWER loading patterns.[13]

8.6. Fast Reactors

The breeder development has been one of the most debatable issues in the research field associated with the nuclear industry. Once nuclear energy gained its share in power production it was already clear that the world uranium resources would be exhausted within a few decades. The estimated breeder benefits and uranium availability have an intimate relationship with each other.[14] It is the expected increase in the uranium price which justifies the larger capital investment inherent to the breeder. As of 1975, the state of the art was characterized by the prototype phase of the liquid moderated fast breeder reactor (LMFBR).

The development front has during the 1970s advanced to the planning of a commercial size (1200 MWe) breeder[15] with a core resembling closely the prototype designs. A remarkable feature of the LMFBR is the high power density achieved. A 3000 MWt core has a height of only 1 m. For the sake of curiosity one should point out that the core height of 1.5 m and the radius of 3 m have been suggested[16] as feasible for a 25,000 MW (10,000 MWe) core utopian at the present time.

The French SUPERPHÉNIX core design[15] employs hexagonal fuel assemblies containing some 250 fuel pins. The pellet O.D. is 7 mm. The core is divided into two radial fuel zones, each of which consists of 150–200 fuel assemblies. Both the short- and long-term reactivity compensation uses control assemblies ordered in two circles. The first ring is located within the inner fuel zone while the second ring follows the interface between the two fuel zones.

The French core is designed for annual refuelling. The radial power distribution across the two core zones is drawn in Fig. 8.5, at BOC and EOC. The flux depression due to the control assemblies is readily detectable in Fig. 8.5. The radial power flattening is adjusted by the initial enrichments of 14.5 and 18.5% in the inner and outer zones, respectively.

The development schedule of the helium-cooled fast breeder (GCFR) follows the LMFBR program and rather detailed suggestions concerning the core management of either the prototype or the full-scale core are available in the literature.[17]

The neutron spectrum present in a typical fast reactor core has

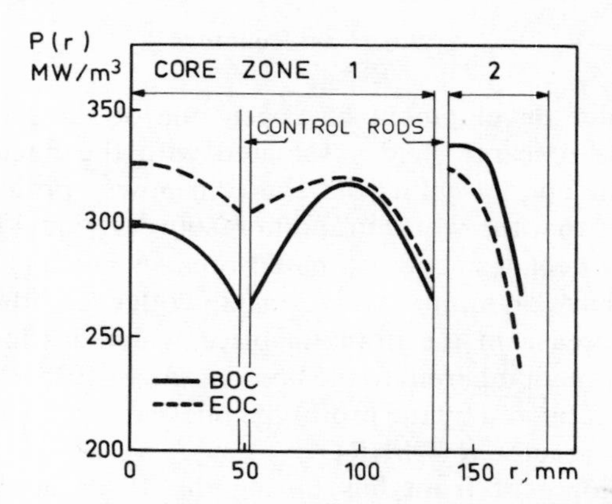

FIG. 8.5. Radial power distribution in a fast core. Reproduced from ref. 15 with permission of Commissariat à l'Energie Atomique.

already been sketched schematically in Fig. 1.9. There are increased complexities introduced by the resonance structure of heavy fuel isotopes and of some light and intermediate mass elements, such as oxygen and iron. The resonance interactions become manifold with emphasized interference and overlapping. Already Fig. 1.9 indicates the rapid depletion of the neutron spectrum at resonance energies. To calculate the sensitive fuel temperature coefficients within the desired accuracy necessitates the use of a fine energy mesh.

The preparation of the energy-averaged and space-homogenized cross-sections is conventionally based on the P_1 or B_1 calculations described in section 7.3. Recall now, however, that hydrogen is no more the most capricious element because of its minor concentration in the material inventories. In comparison with the fast spectrum calculation displayed in section 7.3, the simplifying age or Greuling–Goertzel approximations are no more usable and program packages employ usually a direct lethargy discretization of eqs. (7.2)–(7.3).[18] To demonstrate the technique let the sum over m be carried out and include the inelastic scattering in the scattering kernel in eqs. (7.2)–(7.3). In other words, the procedure is commenced from eqs. (2.18)–(2.21) where lethargy is used as an independent variable. The Fourier transformed equations corresponding to eqs. (7.39)–(7.40) are cast in the form

$$\frac{B_0^2}{|B_0|}\hat{J}(B_0u)+\Sigma_t(u)\hat{\phi}(B_0,u)$$
$$=\int_{-\infty}^{u}\Sigma_s(u')f_{s0}(u',u)\hat{\phi}(B_0,u')\,du'-\hat{Q}_0(B_0,u), \tag{8.4}$$

$$-\frac{|B_0|}{3}\hat{\phi}(B_0,u)+\gamma(B_0,u)\Sigma_t(u)\hat{J}(B_0,u)$$
$$=\int_{-\infty}^{u}\Sigma_s(u')f_{s1}(u',u)\hat{J}(B_0,u')\,du'+\hat{Q}_1(B_0,u) \tag{8.5}$$

where the B_1 function γ was defined in eq. (7.25). The upper limit of integration is placed at u and therefore only down-scattering is considered. B_0 denotes the fundamental mode buckling. The intermediate steps of the derivation are omitted.

Equation (8.4) is discretized exactly in the same manner as eq. (2.44) was derived and one has

$$\frac{B_0}{|B_0|}\hat{J}_i(B_0)+\Sigma_{ti}\hat{\phi}_i(B_0)=\sum_{j=1}^{i}\Sigma_{sji}\hat{\phi}_j(B_0)+\hat{q}_i(B_0). \tag{8.6}$$

If no diffusion coefficient is specified, then one has to accept the current weighting in eq. (8.5). If the first order Legendre component f_{s1} is known as a function of the lethargy, eq. (8.5) can be written in the form

$$-|B_0|\hat{\phi}_i(B_0)+3\gamma_i(B_0)\Sigma_{ti}\hat{J}_i(B_0)=\sum_{j=1}^{i}\Sigma_{sji}^1\hat{J}_j(B_0)+\hat{q}_i^1(B_0). \tag{8.7}$$

Observe that Σ_{ti} is weighted by the flux as normal and therefore the weighting of γ_i

$$\int_{\Delta u_i}\gamma(B_0,u)\Sigma_t(u)\hat{J}(B_0,u)\,du=\gamma_i(B_0)\Sigma_{ti}\hat{J}_i(B_0) \tag{8.8}$$

is not quite straightforward. The first order anisotropic scattering cross-section Σ_{sji}^1 is naturally weighted by the current

$$\Sigma_{sji}^1=3\frac{\int_{\Delta u_i}du\int_{\Delta u_j}\Sigma_s(u')f_{s1}(u',u)\hat{J}(B_0,u')\,du'}{\hat{J}_j(B_0)} \tag{8.9}$$

where the factor 3 is associated with the first Legendre component. $\hat{q}_i^1$ denotes the quantity

$$\hat{q}_i^1 = \int_{\Delta u_i} \hat{Q}_1(B_0, u)\, du. \tag{8.10}$$

While the above scheme is less intuitive than the corresponding one outlined in section 7.3, one must recognize that the need for principal nuclear data is increased in letting the anisotropic components be present explicitly. In terms of condensing the cross-sections into a few group sets, there is no difference relative to the earlier treatment. However, the down-scattering is heavier and the lethargy increment per collision will remove neutrons to lower groups than only the adjacent one.

Due to the lethargy sensitivity full-core fast reactor simulation is conducted in up to ten groups. Diffusion theory is adequate and the finite difference scheme of eqs. (8.2)–(8.3) is applicable. There are even certain three-dimensional fast-reactor simulator programs available in the computer code libraries.[19]

References

1. Deanigi, D. E., *Nuclear Technology*, **18**, 80 (1973).
2. Uotinen, V. O. *et al.*, *Nuclear Technology*, **18**, 115 (1973).
3. Haley, J., WCAP-4167, Westinghouse Electric Corporation, Pittsburgh, Penn., 1969–71.
4. Crowther, R. L. *et al.*, *Trans. Am. Nucl. Soc.* **17**, 297 (1973).
5. Dahlberg, R. C. *et al.*, GA-A12801 (Rev.), General Atomic Company, San Diego, Calif., 1974.
6. Merril, M. H., GA-A1265s, General Atomic Company, San Diego, Calif., 1973.
7. Rippon, S., *Nucl. Engng. Int.* **19**, 659 (1974).
8. Briffs, A. J. *et al.*, *J. Br. Nucl. Energy Soc.* **11**, 215 (1972).
9. Brinkworth, M. J., AEEW-R631, Atomic Energy Establishment, Winfrith, Dorchester, 1969.
10. Page, R. D., *Nucl. Engng. Int.* **19**, 496 (1974).
11. Pasanen, A. A., in *Experience from Operating and Fuelling Nuclear Power Plants*, International Atomic Energy Agency, Vienna, 1974.
12. Denisov, W. P. *et al.*, in *Peaceful Uses of Atomic Energy*, Vol. 2, United Nations, New York and International Atomic Energy Agency, Vienna, 1972.
13. Koskinen, E. and Silvennoinen, P., *Trans. Am. Nucl. Soc.* **20**, 376 (1975).
14. Manne, A. S. and Yu, O. S., *Nuclear News*, **18**, 46 (1975).
15. Clauzon, P. *et al.*, *Bulletin d'Informations Scientifiques et Techniques du Commissariat à l'Energie Atomique*, No. 182, 13 (1973).
16. Kobayashi, Y. *et al.*, *J. Nucl. Sci. Technology*, **10**, 607 (1973).
17. Cerbone, R. J., *J. Br. Nucl. Energy Soc.* **12**, 409 (1973).
18. Archibald, R. J. and Mathews, D. R., GA-7542, Vols. I and II, General Atomic Company, San Diego, Calif., 1968–73.
19. Hardie, R. and Little, W., BNWL-1264, Battelle Memorial Institute, Richland, Wash., 1970.

PART III

OPTIMIZATION AND SYSTEM INTEGRATION

CHAPTER 9

Methods of Optimization

THE computational analysis for reactor core fuel management was commenced in Chapter 7 from the fuel pin level and was carried out to the extent of burnup-dependent core simulation. The next logical step is to study the interface between the core description and the economic objectives assigned for power production. The flow of pertinent information is first studied in the direction from the power production system to the optimization of the fuel costs for a single reactor.

It would not be surprising if someone had suggested in the early days of nuclear reactors that the performance optimization would be a routine matter of applying the existing techniques of mathematical programming and getting the feasible and optimal solution without any particular concern at all. The absolute obstruction one faces, however, is the formidable requirement of computer storage and running time. Mainly for this reason the development of applicable optimization routines is lagging behind the core analysis methods by a decade at least.

It was pointed out in Chapter 7 that the procedures displayed are standard and used all over the world. This statement does not apply to the present chapter, but the methods are under perpetual refinement and practical experimentation. One may only wish that by the late 1970s widely accepted practices will be established in the field. There has been much interest devoted to the short-term utility planning problems and one of the first large symposia was promoted in 1974.[1]

9.1. Fuel Cycle Economics

Proceeding in chronological order, the process of fuel cycle optimization commences from defining the cycle energy E_k to be extracted from the given reactor core during the core cycle k. E_k is dispatched from the utility system model and it is based on the incremental cost of producing the energy over a given planning horizon. The incremental cost is defined as the ratio of the change in revenue requirements per change in the produced amount of energy. The utility naturally aims to minimize the cost under the constraints imposed.

Having other sources of energy than nuclear available, the accumulated operating experience is employed for determining the incremental cost for each combination of the energy sources. In particular, if the nuclear capacity is small the reactor plant will be operated baseloaded at the highest attainable load factor. In Chapter 10 there will be given reasons why this is not always the case and therefore the specification of E_k is more general. It should be emphasized, however, that since the fuel residence time in the core lasts usually at least three cycles, a sequence of cycle energies $E_k, E_{k+1}, \ldots,$ should be given as input. For $l > k$ the E_l's are estimated and later when the cycle l is actuated a more precise value of E_l is determined by the circumstances valid at that time.

Given the required thermal cycle energy E_l, a reference fuel loading is looked for by seeking a feasible combination of the batch size and enrichment. For the BWR where fuel depletion is compensated by control rods, the reference strategy implies the sequencing of the rod movements as well. The reference loading pattern search converges rapidly even if one starts from the idealized out–in scatter loading in Fig. 6.6.

The reference cycle strategy is used as the initial point for optimization routines which are provided for the refinement of the strategy. The final optimal solution is required to yield the minimum of the levelized fuel cycle costs. The concept of levelized cost involves an allocation of the energy costs to consecutive cycles in the proportion that the actual consumption of fuel takes place during the residence time.

A batch cannot always be regarded as the smallest unit of loading or discharge. The fuel assemblies which are loaded at the same time and experience about the same exposure before being discharged at the same refuelling are referred to make up a fuel lot. Let the fuel lot l be loaded at time t_c^l and discharged at t_d^l such that at least part of the interval (t_c^l, t_d^l) coincides with the kth cycle (t_{k-1}, t_k). If $E^l(t)$ denotes the energy produced from the lot l until time t then the total cycle energy output E_k is obtained from

$$E_k = \sum_l [E^l(t_k) - E^l(t_{k-1})] \tag{9.1}$$

where the sum is extended over all the lots present.

The present value of the fuel lot depends naturally on the irradiation status. At t_c when the fuel is loaded its value is determined by $M_U C_U / f_f + M_U C_f$, where f_f is the fractional yield of uranium in the pre-irradiation processes, M_U is the uranium mass in the lot loaded while C_U and C_f are the unit cost of enriched uranium and fabrication plus conversion, respectively. Therefore, C_U consists of both raw uranium and enrichment costs. At any subsequent time prior to the discharge at t_d the value $V_l(t)$ of lot l develops according to[2]

$$\begin{aligned} V_l(t) = {} & M_U(t) C_U(t) + M_{Pu}(t) C_{Pu}(t) \\ & + \left(1 - \frac{E^l(t)}{E^l(t_d^l)}\right)\left[\left(\frac{1}{f_f} - 1\right) C_U(t_c^l) + C_f(t_c^l)\right] M_U(t_c^l) \\ & - \frac{E^l(t)}{E^l(t_d^l)} [(1 - f_r) M_{Pu}(t_d^l) C_{Pu}(t_d^l) \\ & + f_r C_c(t_d^l) M_U(t_d^l) + (1 - f_r f_c) C_U(t_d^l) M_U(t_d^l) \\ & + C_s(t_d^l) M_U(t_c^l) + C_r(t) M_U(t_c^l)]. \end{aligned} \tag{9.2}$$

In eq. (9.2) C_c, C_s and C_r denote the unit charges of post-core conversion, shipment and reprocessing, respectively. Since the prices are functions of time, they must be estimated for any future action. f_c and f_r denote the material yields in conversion and reprocessing, respectively. The factor $E^l(t)/E^l(t_d^l)$ proportions only the values with respect to the exposure. In particular, $E^l(t_d^l)$ is estimated on the basis of core simulation.

Equation (9.2) is merely an extension of the simplified value analysis of section 6.6. Now the yields f_i have been introduced in order to account properly for the minor material losses occurring during fuel processing. The discharge fuel is assumed to be reused for the portion during which uranium is present and for plutonium the estimated effective value C_{Pu} is to be used.

The levelized costs for the cycle k are determined as the

expenditures associated with the fuel lot values per the amount of energy produced when both quantities are allocated over consecutive cycles in the real proportion. To account for the coupling between consecutive cycles, the fuel inventory value must be calculated at times outside the interval (t_{k-1}, t_k). For example,[2] one may choose times t_1^k and t_2^k in the middle of the cycles $k-1$ and $k+1$,

$$t_1^k = t_{k-1} - \tfrac{1}{2}(t_{k-1} - t_{k-2}) \tag{9.3}$$

$$t_2^k = t_k + \tfrac{1}{2}(t_{k+1} - t_k). \tag{9.4}$$

The direct revenue requirement for lot l over the period (t_1^k, t_2^k) is obtained from

$$R_l(t_0, t_1^k, t_2^k) = V_l(t_1^k)/(1+i)^{t_0 - t_1^k} - V_l(t_2^k)/(1+i)^{t_0 - t_2^k} \tag{9.5}$$

where the expenditures are discounted to t_0. i denotes the rate of interest or the effective cost of money. No allowance is made for taxation or depreciation deductions in eq. (9.5), but it would be a simple matter to modify the revenue requirement R_l according to the rules that are valid for any particular reactor operator. The carrying charges for the periods prior to the lot loading or after the lot discharge must be discounted to t_0. The corresponding revenue requirement is denoted by $R_l^c(t_0, t_1^k, t_2^k)$ which is already proportioned by the factor $(E^l(t_2^k) - E^l(t_1^k))/E^l(t_d^l)$ from the total out-of-core carrying charges for lot l and discounted to t_0.

The value function $V_l(t)$ in eq. (9.2) was written in the form where the lot dwell time and total energy from the lot were estimated. Also some of the unit costs include uncertainties due to developing cost trends. Within the error caused by these factors, the levelized costs for cycle k can be expressed as

$$c_k = \frac{\sum_l [R_l(t_0, t_1^k, t_2^k) + R_l^c(t_0, t_1^k, t_2^k)]}{\sum_l [E^l(t_2^k) - E^l(t_1^k)](1+i)^{t_1 - t_0}} \tag{9.6}$$

where the energy production is also discounted to the same time t_0. t_1 denotes the time when revenue is received for the energy generated from the lot l. In case the cash flow for energy were based on shorter subintervals $(s_0, s_1), (s_1, s_2), (s_m, s_{m+1}) \ldots (s_{M-1}, s_M)$ then an immediate correction should be made in eq. (9.6) which would have the form

$$c_k(t_0, t_1^k, t_2^k) = \frac{\sum_l [R_l(t_0, t_1^k, t_2^k) + R_l^c(t_0, t_1^k, t_2^k)]}{\sum_{m=1}^{M} \sum_l [E^l(r_m) - E^l(r_{m-1})](1+i)^{s_m - t_0}} \qquad (9.7)$$

where r_m is the end of period m for which the payment is received at s_m. Hence $s_m - r_m$ is the lag time involved. Clearly $r_0 = t_1$ and $r_M = t_M$.

To recapitulate the essence of eq. (9.7), the levelizing interval (t_1^k, t_2^k) is extended to a wider period than the cycle length (t_{k-1}, t_k). The cycle energies from each lot l are estimated. In case it will be verified at a later time that there was any discrepancy involved from what actually was generated, eq. (9.7) is subject to a correction for the part of the discharged lots. Only by making the adjustment is one able to end up with the actual total revenue requirement that was needed for a longer planning horizon than (t_1^k, t_2^k).

The minimization of c_k must be conducted in a manner which involves a proper consideration of the reactor state. At no time during the cycle can the power peaking exceed the upper limit specified. The exit burnup τ_k must be confined to a value that guarantees undisturbed fuel performance.

Recalling that the actual process of optimization involves the loading pattern, let $\mathbf{x}(k)$ denote the vector that contains the information of assembly positions and exposures. Physically, it includes the distribution of fissile and poison isotopes within the core. At each refuelling a decision must be made on the fraction α_k of fuel to be discharged.

A further decision variable is ϵ_k, the initial enrichment of the reload batch. Finally, one has the control vector $\mathbf{y}(k)$ which incorporates all the adjustable control measures, i.e. control rod sequences and soluble poison concentrations. The objective function can then be formally given by

$$\min c_k = \min c_k[\mathbf{x}(n), \mathbf{y}(n), \alpha_n, \epsilon_n] \qquad (9.8)$$

over all the cycles n within the planning horizon

$$(t_{n-1}, t_n) \subset (t_1^k, t_2^k). \qquad (9.9)$$

While eqs. (9.8)–(9.9) state the achievement of optimum for the cycle k, it is clear that the coupling is retained to the other cycles whether preceding or succeeding ones.

Equation (9.8) is subject to the necessary constraints

$$\tau_n < \tau_{\max}, \tag{9.10}$$

$$F_R F_Z < (F_R F_Z)_{\max} \tag{9.11}$$

where $\tau_{\max}$ denotes the maximum permissible exit burnup of fuel and $(F_R F_Z)_{\max}$ denotes the limiting peaking factor. In order for any solution $\{\mathbf{x}(n), \mathbf{y}(n), \alpha_n, \epsilon_n\}$ to be feasible it has to yield the desired cycle energy E_n with a specified cycle length (t_{n-1}, t_n).

9.2. Loading Pattern Search

Once the sequence $\{E_k\}$ of the specified thermal energies during consecutive cycles is given, the planning activities involve the generation of feasible alternative loading and shuffling patterns. This is needed for the purpose of providing input to the optimization modules. Since refuelling can take place at less frequent timesteps than what the control vector $\mathbf{y}(n)$ can be modified, it is reasonable to omit $\mathbf{y}(n)$ for a while and express the levelized costs c in the form

$$c = c[\mathbf{x}(n), \alpha_n, \epsilon_n]. \tag{9.12}$$

Recalling that the PWR, for example, employs homogeneous reactivity compensation by means of soluble poisons, eq. (9.12) is valid for the entire optimization in this case.

Since the loading pattern in the LWR will in the first approximation be given by the idealized out–in scatter loading in Fig. 6.6, it is quite easy to condense the information contained in $\mathbf{x}$. For the inner core region where the fraction $(1-\alpha_k)$ of the fuel is located, the average exposure τ_k^{in} and the average fissile content ϵ_k^{in} are adequate parameters.[3] ϵ_k^{in} denotes the ratio of U^{235}, Pu^{239} and Pu^{241} mass divided by the total mass of uranium and plutonium. The precalculated correlations to specify an approximative value of the reload batch enrichment ϵ_k are of the form

$$\alpha_k \epsilon_k = \alpha_k \epsilon_k (E_k, \tau_k^{\text{in}}, \epsilon_k^{\text{in}}). \tag{9.13}$$

The functional form of eq. (9.13) is established in terms of polynomials

$$\alpha_k \epsilon_k = \sum_l \sum_m \sum_n a_{lmn} (E_k)^l (\tau_k^{\text{in}})^m (\epsilon_k^{\text{in}})^n \tag{9.14}$$

where no higher than quadratic terms are usually required.[3] The coefficients a_{lmn} are obtained from an earlier history of operation and they correspond to each specific core design.

An evaluation of the EOC state of the core must be performed to make sure that no excessive burnups will be involved. Inverting the relation in eq. (6.17) one obtains for the exit burnup

$$\tau_k = b_0 + b_1 \epsilon_k \tag{9.15}$$

which facilitates the first screening of nonadmissible discharge burnups. The coefficients b_0 and b_1 are again design sensitive.

The next degree of sophistication would imply use of a diffusion theory core description combined with formal mathematical programming already at the stage of generating the loading alternatives. An interesting example along these lines is given in the Norwegian FALC code[4] which includes a radially multizone treatment of the core and a standard linear programming module. Combining FALC with a one-dimensional survey code, average power densities across the zones are computed over a number of fuel cycles and the feasible optimal k_∞ is solved regionwise. A phase of the calculation is summarized in Fig. 9.1.

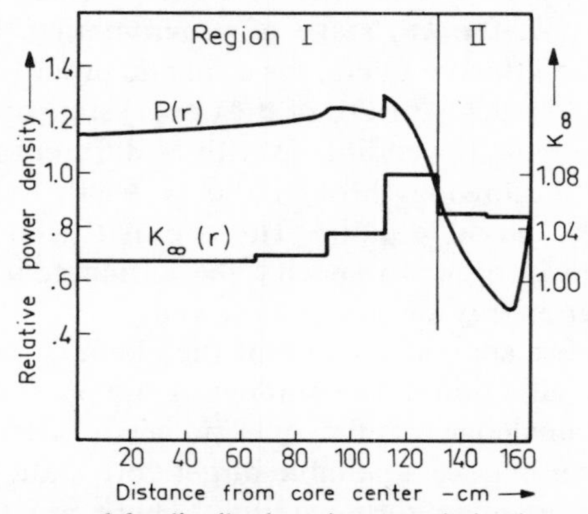

FIG. 9.1. Radial power and k_∞ distributions in an optimization study. Reproduced from ref. 4 with permission of the International Atomic Energy Agency.

The core is divided radially into two macrozones with an additional subdivision for the one-dimensional mesh. The k_∞-distribution is closely linked to the ϵ_k as well as ϵ_k^{in} values. The sample problem in Fig. 9.1 is run for a BWR,[4] while the loading pattern search is even more relevant for PWRs for the reasons stated earlier.

Having a number of feasible loading strategies available, one is able to start the actual search for the optimal patterns. This phase is mostly relevant for the PWR because the BWR does not allow for the continuing omission of the control vector **y**. When coupled to intuition, the techniques of dynamic programming have shown noticeable promise in the loading pattern search and the following discussion is based on this programming mode. One should recognize, however, that future development may turn out in favour of some other approach. A brief summary of the other methods appears in ref. 5 in a form readable for reactor core analysts.

A dynamic programming solution of the type presented in ref. 3 will be outlined below. The three batch out–in scatter loading for the PWR implies a batch fraction $\alpha = 1/3$ in steady state. If α_k is selected to be a decision variable and the reactor is operated close to the ideal performance limit with annual refuelling and with a load factor between 80 and 90%, then it is natural that α_k must vary around 0.33 for all cycles. Assume that at each refuelling decision one has N different values of α_k available, then core history over the cycle k can be described by an index P_n where $1 \leq n \leq N$. All possible sequences of refuelling decisions over the planning horizon span a network graph. At the kth stage of programming, i.e. concerning decisions for the kth core cycle, the dynamic programming problem is illustrated in Fig. 9.2. At EOC of the $(k-1)$ state one may proceed along any of paths $P_n(k)$ ending up with N different EOC states for the kth cycle. Following through the $(k+1)$th cycle one would already have N^2 possible paths. The use of the correlation in eq. (9.14) is needed in order to specify the enrichment ϵ_k so that the desired thermal energy output is achieved.

In case the core state at the end of the planning horizon is fixed, there occurs a reduction in the number of the possible paths. Since there is no knowledge of the energy needs after the planning interval, one can impose a suitable target core state which must be reached within a given time interval. There are two immediate

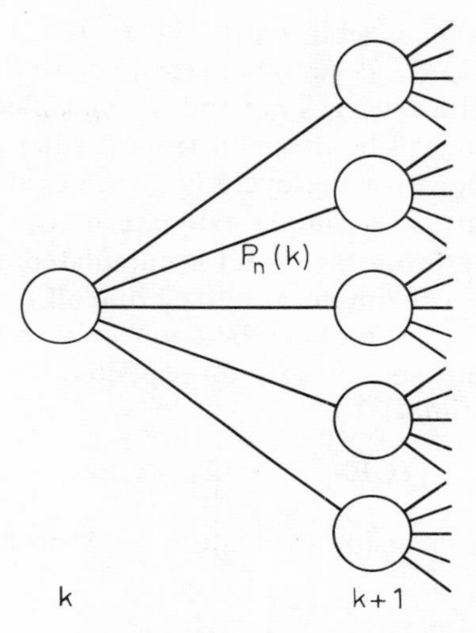

FIG. 9.2. Decision tree.

propositions concerning the target state variables. Firstly, they can be set to correspond to an equilibrium situation with a constant cycle energy and $\alpha = 1/3$. Secondly, the target state can be selected to be the one prevailing at the beginning of the planning horizon.

At the stage k a further reduction of the paths can be accomplished by means of the exit burnup check in eq. (9.15). Let the sequences $P_l(1) - P_m(2) - \ldots - P_n(k-1)$ with variable indices l, m, n, etc., denote the core histories up to BOC of the stage k. Applying educated judgement on the total fuel cycle cost incurred, one is able to eliminate the economically diverging paths for the cycle k.

One is not necessarily interested in the complicated revenue allocation principle of eq. (9.7), but the accumulated total revenue requirement is well suitable for the economic bookkeeping. Over the entire planning horizon the two criteria should agree, of course. Rather than use the differential revenue requirement as in eq. (9.5), one can follow the total expenditures C_k comprising the direct as well as carrying charges for the fuel lots procured until and including the cycle k. The C_k's are already assumed to be discounted to a given time.

Along any of the feasible paths $P_l(1) - P_m(2) - \ldots$ the accumulated total cost $\Sigma_{i=1}^{k} C_{i,p}$ is computed. p specifies the path concerned. As soon as the path diverges beyond the specified margin from the latest best path, it will be dropped from further considerations. In making the decisions for the cycle k by the evaluation of different paths, one should use a simple estimate of the effect exerted on future cycles. Therefore the direct accumulated cost must again be augmented by a levelizing term to describe all the feasible progeny $\{P(k+1), P(k+2) \ldots\}$ that can be connected to p. Considering the stage $k+1$ only one searches for an expected cost estimate $CE_{k+1,p}$. The total cost estimate[3]

$$\mathrm{TCE}_{k,p} = \sum_{i=1}^{k} C_{i,p} + \mathrm{CE}_{k+1,p} \tag{9.16}$$

is a more reliable decision evaluation function than the total cost

$$\mathrm{TC}_{k,p} = \sum_{i=1}^{k} C_{i,p} \tag{9.17}$$

would be.

Without introducing the exact quantitative parameter values, the solution of a sample problem is illustrated in Fig. 9.3. Kearney *et al.*[3] have applied the dynamic programming procedure to a 1050 MWe PWR and solved the optimal fuel policy over a planning horizon of five years. Both the cycle energies and cycle durations have varied. A total of four discrete batch fractions are used as decision variables with the options $A = P_1$, $B = P_2$, $C = P_3$ and $D = P_4$ corresponding to 0.37, 0.33, 0.29 and 0.25, respectively. The total discounted revenue requirement for the optimal policy C-A-D-D-C is denoted by TC* in Fig. 9.3. The paths are terminated either due to excessive discharge burnup (x) or due to excessive cost (o). After the proper planning interval the core is returned to its initial state during the four lag cycles. The relative savings can be deduced from Fig. 9.3 on the basis that TC* = $\$51.3 \times 10^6$.

The computations pertaining to Fig. 9.3 have been performed under the assumption of constant unit costs. It is worth while to emphasize, however, that the explicit time dependence of the value function in eq. (9.2) should not be ignored in the circumstances where anticipated cost trends are available.

Clearly the steady-state planning is a special case of the procedure

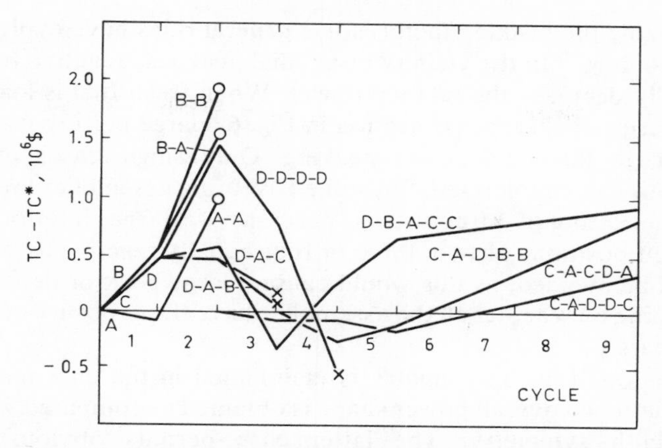

FIG. 9.3. Deviation of fuel strategies from the optimal policy.[3]

described. After the equilibrium state is achieved, the levelized cycle cost and the total revenue requirement are functions of the constant initial enrichment and batch fraction

$$c = c[\alpha, \epsilon]. \tag{9.18}$$

The solution of the loading pattern problem has been studied for the PWR optimizing the burnup or minimizing the radial power peaking factor as the objective function. Examples of these approaches appear in refs. 6 and 7, respectively. Power flattening is an attractive objective in the circumstances where the energy output of the plant is limited by excessive power peaking. Frequently, however, the core energy output is not the only restriction, but the other parts of the plant, e.g. turbines, prohibit an increase in power and the economic gain is not accessible.

Reducing the power peaking factor during a single fuel cycle exerts a minor coupling to the other cycles and hence the planning period can be divided more easily into single decision stages than is the case in the cost levelizing method. To account for the economic advantages the minimization of the peaking factor should be coupled to a simultaneous search based on financial grounds.[8]

Independently of whether the calculation of the power shape is aimed at checking the limit in eq. (9.11) or at the objective of minimizing the peaking factor, some general rules have evolved for fuel shuffling.[7] In the vicinity of a radial peak less reactive fuel will naturally decrease the relative power. When fresh fuel is loaded in the interior checkerboard section in Fig. 6.6 large perturbations are caused on the local power peaking. Only single cases of such locations can be tolerated. Since the PWR fuel assemblies are large, the checkerboard structure is necessary. In the interior core, adjacent positions of two, three or four equally reactive assemblies should be avoided, as this would cause a large peak or depression, depending on whether all the assemblies have the exposure of one or two years.

If the one-fourth symmetry is maintained in the core loading a more flattened overall power shape is obtained as compared with the one-eighth symmetry. The latter case permits obviously less flexibility.

The rotating of irradiated fuel assemblies must be done intuitively, because the reactor simulator treats homogenized assemblies only. The rotations are relevant for those assemblies which are moved from the core periphery into the checkerboard zone after their first in-core cycle. Given the core position the rotation should result in a situation where the most reactive subassembly is faced by the most depleted adjacent assembly.

9.3. Control Rod Programming

Practically no concern has yet been given to the optimization of the axial power and burnup distributions. In the regular base load mode the PWR is controlled homogeneously and deserves no further discussion in this respect. On the other hand, depletion in the BWR core is compensated by withdrawal of the control rods which leaves numerous degrees of freedom available. Control rod programming refers to the optimization of the rod withdrawal in the sense that specified performance criteria will be implemented.

Consider first the BWR control rod withdrawal as a problem to be solved by a mathematical programming procedure. Since the axial flux shape is rather nonuniform, the maximum energy output is to a large extent confined by the axial peaking or by the imminence of surpassing the MCHFR. To yield the desired cycle energies a smooth axial power shape is also most economic because it maximizes the exit burnup and therefore minimizes the reload enrichment.

It is assumed here that the loading pattern search has produced close-to-optimal batch fractions and enrichments and therefore the rod programming state concerns a cost function

$$c = c[\mathbf{x}(n), \mathbf{y}(n)]. \tag{9.19}$$

Differently from eq. (9.8), n denotes now a single stage which is identical not to one cycle but only to a subinterval thereof. To formulate the objective function in terms of minimizing the cycle costs c_k along the stage graph would be rather academic since practical optimization algorithms employ the power shaping as the performance criterion. Let $F(n)$ denote the core peaking factor at the nth stage during the core cycle. If Y denotes the set of control decisions $\{y(1), y(2) \ldots\}$ then the optimal policy is associated with the index I,[9]

$$I = \min_{\substack{\text{over} \\ \text{all } y}} \max_{\substack{\text{over} \\ \text{all } n}} F(n). \tag{9.20}$$

Commencing from the initial state of the core at BOC the application of eq. (9.20) generates a network of stage-to-stage paths which at EOC define the reachable domain of discharge core states.

The minimax criterion in eq. (9.20) concerns the global power peaking only. If one would attempt to minimize the local power variations a reference axial power distribution $P_0^m(z)$ should be employed. m refers to the iterative nature of the problem where $P_0^m(z)$ is updated when desired. The optimum would be reached by the requirement

$$I^m = \min_{\substack{\text{over} \\ \text{all } Y}} \max_{\substack{\text{over} \\ \text{all } k}} \int_0^Z |P(z,k) - P_0^m(z)| \, dz \tag{9.21}$$

where Z denotes the core height. A new iterate P_0^{m+1} can be constructed from the solution $P(z, k)$.

Equation (9.21) actually includes the Haling principle introduced in section 6.2. $P_o^m = P_0$ is chosen to have the form of the EOC exposure distribution and the criterion is written by[9]

$$I = \min_{\substack{\text{over} \\ \text{all } Y}} \max_{\substack{\text{over} \\ \text{all } k}} \int_0^Z |P(z,k) - P_0(z)|^2 \, dz. \tag{9.22}$$

Computations performed in ref. 9 indicate the expected trend that eq. (9.20) yields a somewhat lower global power peaking than eq. (9.22) does. The criterion for I can be modified by means of introducing free coefficients, or Lagrange multipliers, such that negative withdrawal action could be obtained as a result from the minimax study. That would clearly correspond to a rod insertion.

It is prohibitive to determine the power distribution $P(\mathbf{r})$ accurately in a control rod optimization run. Therefore, no full-core analyses are performed, but rather the one-dimensional diffusion equation (7.223) is solved in the axial direction. This means that the core is assumed to be an infinite lattice and therefore corresponds to a scatter loaded case. The important axial effects due to the rods and void fraction are regarded via cross-sections and the radial buckling.

The BWR flux shaping problem is sometimes treated[10] on the Haling principle basis without involving any minimization of residuals. The method can be discussed conveniently within the two-group formalism in eq. (7.205) with all fission neutrons injected in the fast group ($X_1 = 1$) or equivalently one may start from the 1.5 group equation (7.248). Introducing the reactor power $P(\mathbf{r})$ from eq. (1.28)

$$P(\mathbf{r}) = e \sum_n \Sigma_{fn}\phi_n \tag{9.23}$$

where e denotes the energy release per fission (cf. Table 1.2), the fast group or the 1.5 group diffusion equation is now given by

$$-\nabla \cdot D_1(\mathbf{r})\nabla\phi_1(\mathbf{r}) + \Sigma_{r1}(\mathbf{r})\phi_1(\mathbf{r}) = \frac{\lambda\nu}{e} P(\mathbf{r}). \tag{9.24}$$

If the core obeys the Haling principle, then the power distribution must have the form of the EOC exposure distribution $\tau(\mathbf{r})$. An initial guess or a separately optimized solution $\tau(\mathbf{r})$ makes the right-hand side of eq. (9.24) to be known, because the average power P and EOC burnup τ are related to each other:

$$\frac{P(r)}{P} = \frac{\tau(r)}{\tau}. \tag{9.25}$$

Equation (9.24) is reduced to a source problem that can be solved by normal iteration techniques. Note that D_1 and Σ_{r1} can be calculated as soon as the power level, i.e. system temperatures, moderator density and xenon distribution, is known. Once the flux shape $\phi_1(\mathbf{r})$

is solved, a new source term is computed from eq. (9.23) and eq. (9.24) is inverted again until the power shape has converged.[10] The calculations are performed assuming a fixed value of λ, e.g. a critical reactor $\lambda = 1$.

To conclude this section, a few remarks will be made concerning the PWR where power manoeuvring introduces the need to program the control rod movements. The partial length rod facility is of primary importance. For the PWR the rod programming is mainly a question of reactor control rather than of fuel optimization.

Recalling the xenon characteristics discussed in section 4.5, control rod insertion may cause severe power peaking and axial oscillations in the PWR. Since I^{135} has a half-life of 6.7 hr, the xenon effects are associated with control rod insertions of a few hours' duration. Figure 9.4[11] depicts a situation where the power of a 1000 MWe PWR is reduced to 60% for a period of 8 hr. The axial flux profile is drawn in Fig. 9.4 prior to the rod insertion and immediately after the withdrawal.

After the rods are withdrawn there occurs a sizeable power peak in

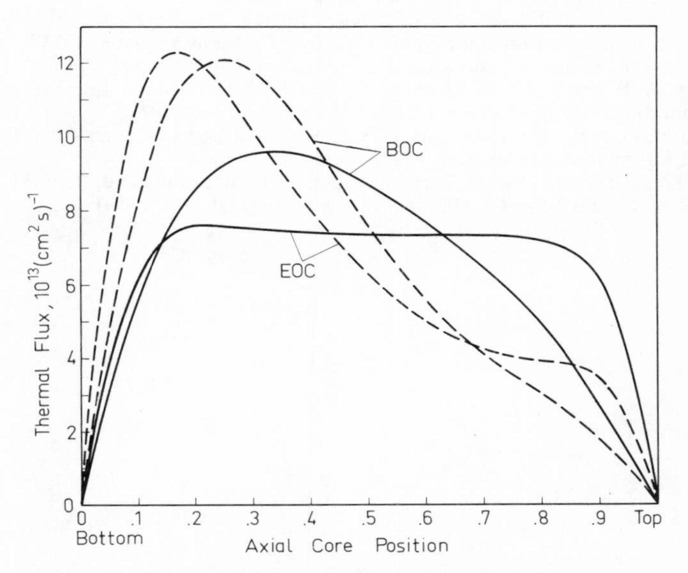

FIG. 9.4. Axial power peaking due to xenon buildup.

the bottom part of the core independently of the burnup status of the core. The 8 hr duration of the power cutback is typical in daily load following (cf. Fig. 10.2). Since the axial peaking factor reached 1.9, this mode of operation would not be licensable at full power. The partial length rods can, however, be positioned at the height at which the peak occurs and they can be relocated in case the peak position begins to oscillate. The rod withdrawal can be executed in a stepwise manner to smoothen the transient.[12]

References

1. *A Collection of Papers Presented at the Nuclear Utilities Planning Methods Symposium*, ORNL-TM-4443 (Rev.), Oak Ridge National Laboratory, Oak Ridge, Tenn., 1974.
2. Rieck, T. A. *et al.*, article included in ref. 1.
3. Kearney, J. P. *et al.*, article included in ref. 1.
4. Sauar, T. O. *et al.*, in *Peaceful Uses of Atomic Energy*, Vol. 2, United Nations, New York, and International Atomic Energy Agency, Vienna, 1972.
5. Sesonske, A., *Nuclear Power Plant Design Analysis*, U.S. Energy Commission, Technical Information Center, Oak Ridge, Tenn., 1973.
6. Stover, R. L. and Sesonske, A., *J. Nucl. Energy*, **23**, 679 (1969).
7. Stout, R. B. and Robinsson, A. H., *Nuclear Technology*, **20**, 86 (1973).
8. Koskinen, E. and Silvennoinen, P., *Trans. Am. Nucl. Soc.* **20**, 376 (1975).
9. Snyder, B. and Lewis, E. E., in CONF-730414-P1, U.S. Atomic Energy Commission, Technical Information Center, Oak Ridge, Tenn., 1973.
10. Crowther, R. L., in CONF-730414-P1, U.S. Atomic Energy Commission, Technical Information Center, Oak Ridge, Tenn., 1973.
11. Tiihonen, O., VTT-Ydi-13, Technical Research Centre of Finland, Helsinki, 1974.
12. Bauer, D. and Poncelet, C., *Nuclear Technology*, **21**, 165 (1974).

CHAPTER 10

General System Aspects

THIS last chapter is aimed at illustrating the wider power system considerations in which the problems of reactor core fuel management are embedded. The treatment is rather cursory and concentrates on selected aspects only. The fuel analyst will not customarily work within the environment of the more general system aspects, but certainly it will be of greater assistance to familiarize oneself with the perspective.[1]

10.1. Grid Requirements

Nuclear units possess inherently the feature of involving large plant sizes until the power generation becomes economically competitive. This is why the attraction of nuclear power has been limited to the industrialized part of the world. Even there it has been practical to first introduce small or medium size plants before letting the generation of the 1000 MWe reactors take over.

The primary question in integrating the nuclear power plant into the system concerns the maximum ratio of the total generating output that a single unit can represent. The grid may become unstable if a reactor plant is unexpectedly brought to shutdown. The sudden decrease in power output with constant demand may slow the grid frequency. The other generating plants are not capable of feeding the system, thus leading to a collapse of the distribution network. The behaviour of the other plants depends on the frequency tolerance that they can sustain. Typically a plant corresponding to only 10–15% of the total capacity can induce the effect.

A smaller and more normal power variation in the grid is shown in Fig. 10.1.

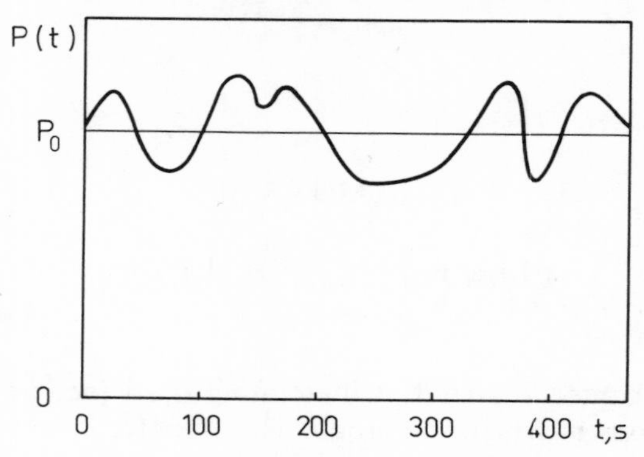

FIG. 10.1. Fluctuations of power demand.

The power fluctuations are caused by stochastic changes in demand. For the British grid, for example, a frequency scale from 49.8 to 50.2 Hz where power compensation is needed has been reported.[2] In this case the variations are very fast and generally one attempts to avoid the frequency control by the reactor.

Since the PWR has a separate steam generator the deviations of the order of 5% per 30 sec can be controlled by means of the large latent heat capacity available (cf. Fig. 5.2). The direct cycle BWR employs the flow rate change, yielding a somewhat weaker fast response.

The grid regulation and stability are problems which have to be solved prior to the purchase or commencement of operation of a plant. The utility has to consider them in the system expansion phase. In the sense in which fuel management is defined in this book, the grid generating capacity also exerts its influence on fuel cycle planning. The overall grid capacity is of utmost importance in scheduling the refuelling outages.

The large nuclear plants must be refuelled during a period when the consumption of electricity is low. Typically society works in such a manner that makes the shutdown undesirable prior to the seasonal holidays in winter and during the expected heavy house heating periods. In contrast, the ventilation and air conditioning systems can cause a load peak in the summertime.

A second contributing factor pertains to the availability of replacing energy and, indeed, the time variation of the replacement energy cost. A splendid example is the hydro-power in countries where the snow melts in the spring and makes the replacement cost very small. There is an incentive to reload early in the summer when the holiday season has simultaneously reduced some of the industrial power load.

10.2. Load Following

Owing to the low fuel-to-investment cost ratio, nuclear power plants are mostly operated constantly at full power to cover the base load requirement of the network. Ignoring the temporary random failures in the conventional fossil-fired plants assigned to load following a nuclear reactor plant may also have to possess a regular capability to follow the system load demand. Suppose, for example, that the nuclear capacity of a utility exceeds the minimum system power demand with the variations depicted in Fig. 10.2.

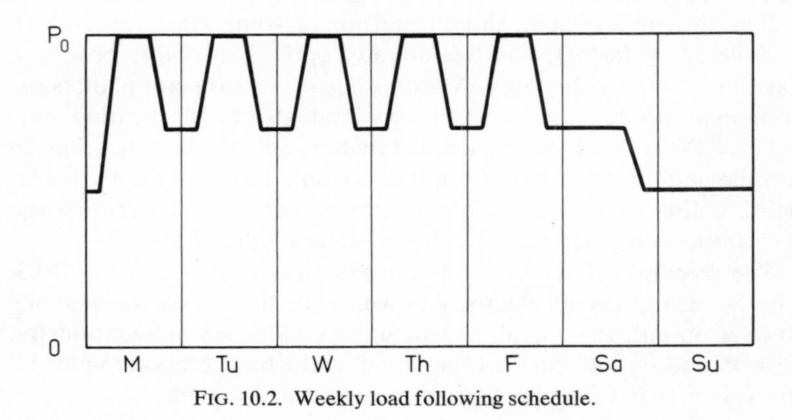

FIG. 10.2. Weekly load following schedule.

Figure 10.2 combines a daily and weekly variation in the reactor power licensed to P_0. There may appear some seasonal aspects as well.

To satisfy the load demand drawn in Fig. 10.2 the core reactivity has to be controlled properly. As far as the PWR is concerned, the control rods are inserted for the cutback period. Their withdrawal has to be reconciled with the xenon buildup effects as discussed at the end of section 9.3. The control rod insertion is, of course, available in all reactors, while some designs including the BWR do not resort to it. In section 4.4 it was indicated that reactivity control through neutron moderation changes is more practical. Due to the negative and dominant moderator temperature, coefficient flow rate changes, i.e. variations in the speed of circulation pumps, furnish the desired response.

10.3. Multipurpose Applications

After nuclear power became commercial, most reactors, especially the large ones, were built for the purpose of providing the steam to produce electricity. While reactors will retain this utility, it is becoming increasingly evident that further uses are competitive in diverse applications. Without resorting to guesswork on whether it will be the process heat, ship propulsion or some other form which is developing fastest, one has already options available based on existing reactor technology. A promising use of a reactor plant is the combined production of electricity and steam for heating purposes.[3] What is technically needed there is only the modifications in turbine unit where a part of the steam would be condensed, the rest being delivered into the district heating network. To reduce the distribution cost necessitates urban siting of the plant.

The principle of the combined production is shown in Fig. 10.3. The y-axis represents electrical power while the x-axis corresponds to heat output. In case all the steam is condensed conventionally, $P_e = P_0$ and $P_{th} = 0$. On the other hand, in the back pressure mode all the steam is fed into heating system $P_e = 0$, $P_{th} = P_h$.

In an intermediate mode where the amount P_1 is extracted into heating, the rest of the output P_2 can be converted into electricity.

Besides the direct economic impetus, the combined production reduces the amount of thermal energy that goes wasted in the condensing process. In view of the optimization procedures of

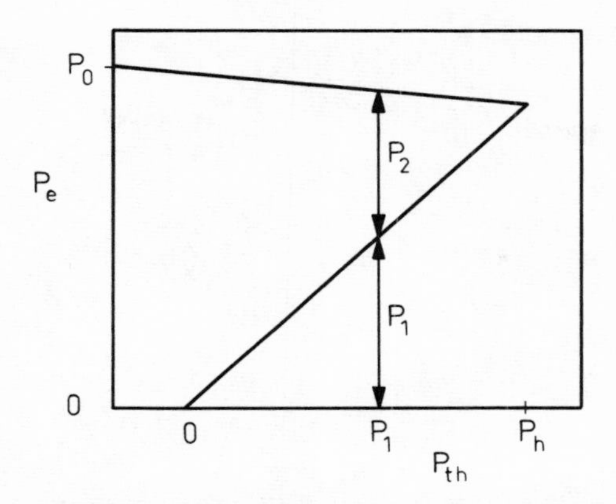

FIG. 10.3. Combined condensing and back pressure operation.

Chapter 9, an additional consideration has to be devoted to the cost entailed in compensating the reduced electricity output. The ratio P_1/P_2 enters the calculation as a key decision variable which depends on heat transmission costs and required annual load duration.[3]

References

1. *Modern Power Station Practice*, 2nd edition, Vol. 8, *Nuclear Power Generation*, Pergamon Press, Oxford, 1971.
2. Rutterfield, M. H., in *Nuclear Power Plant Control and Instrumentation*, International Atomic Energy Agency, Vienna, 1973.
3. Halzl, J. *et al.*, in *Economic Integration of Nuclear Power Stations in Electric Power Systems*, International Atomic Energy Agency, Vienna, 1971.
4. Mikola, J. *et al.*, 9th World Energy Conference, Detroit, 1974.

Index